T0243072

CAMBRIDGE LIBRARY COLLECTION

Books of enduring scholarly value

Technology

The focus of this series is engineering, broadly construed. It covers technological innovation from a range of periods and cultures, but centres on the technological achievements of the industrial era in the West, particularly in the nineteenth century, as understood by their contemporaries. Infrastructure is one major focus, covering the building of railways and canals, bridges and tunnels, land drainage, the laying of submarine cables, and the construction of docks and lighthouses. Other key topics include developments in industrial and manufacturing fields such as mining technology, the production of iron and steel, the use of steam power, and chemical processes such as photography and textile dyes.

Sketch of the Civil Engineering of North America

A distinguished civil engineer, David Stevenson (1815–86) continued his father's work of designing and building lighthouses around the coast of his native Scotland. His three-month tour of the United States and Canada in 1837 resulted in this highly detailed and unprecedented survey, first published in 1838. Stevenson covers a large number of engineering works, ranging from lighthouses and canals through to roads, bridges and railways. Notably, Stevenson's praise for North America's faster and sleeker steam vessels led British shipbuilders to emulate the models he describes and illustrates in this text. The work remains a historically valuable assessment of the continent's infrastructure at a time of great industrial expansion. Stevenson's *The Principles and Practice of Canal and River Engineering* (2nd edition, 1872) and his *Life of Robert Stevenson* (1878), a biography of his father, are also reissued in this series.

Cambridge University Press has long been a pioneer in the reissuing of out-of-print titles from its own backlist, producing digital reprints of books that are still sought after by scholars and students but could not be reprinted economically using traditional technology. The Cambridge Library Collection extends this activity to a wider range of books which are still of importance to researchers and professionals, either for the source material they contain, or as landmarks in the history of their academic discipline.

Drawing from the world-renowned collections in the Cambridge University Library and other partner libraries, and guided by the advice of experts in each subject area, Cambridge University Press is using state-of-the-art scanning machines in its own Printing House to capture the content of each book selected for inclusion. The files are processed to give a consistently clear, crisp image, and the books finished to the high quality standard for which the Press is recognised around the world. The latest print-on-demand technology ensures that the books will remain available indefinitely, and that orders for single or multiple copies can quickly be supplied.

The Cambridge Library Collection brings back to life books of enduring scholarly value (including out-of-copyright works originally issued by other publishers) across a wide range of disciplines in the humanities and social sciences and in science and technology.

Sketch of the Civil Engineering of North America

Comprising Remarks on the Harbours, River and Lake Navigation, Lighthouses, Steam-Navigation, Water-Works, Canals, Roads, Railways, Bridges, and Other Works in that Country

David Stevenson

CAMBRIDGE
UNIVERSITY PRESS

University Printing House, Cambridge, CB2 8BS, United Kingdom

Cambridge University Press is part of the University of Cambridge.
It furthers the University's mission by disseminating knowledge in the pursuit of education, learning and research at the highest international levels of excellence.

www.cambridge.org
Information on this title: www.cambridge.org/9781108071963

This edition first published 1838
This digitally printed version 2014

ISBN 978-1-108-07196-3 Paperback

The material originally positioned here is too large for reproduction in this reissue. A PDF can be downloaded from the web address given on page iv of this book, by clicking on 'Resources Available'.

SKETCH

OF THE

CIVIL ENGINEERING

OF

NORTH AMERICA;

COMPRISING REMARKS ON THE

HARBOURS, RIVER AND LAKE NAVIGATION, LIGHTHOUSES, STEAM-NAVIGATION, WATER-WORKS, CANALS, ROADS, RAILWAYS, BRIDGES, AND OTHER WORKS IN THAT COUNTRY.

BY

DAVID STEVENSON,

CIVIL ENGINEER.

LONDON:

JOHN WEALE, ARCHITECTURAL LIBRARY,

59. HIGH HOLBORN.

MDCCCXXXVIII.

PREFACE

Having at various times heard much to interest and surprise me respecting the engineering works of America, and having been unable to meet with any publication containing satisfactory information regarding them, I resolved to take advantage of a short interval of professional leisure, to examine the subject for myself.

In a tour of about three months I visited Upper and Lower Canada, and the most interesting parts of the United States of America, and endeavoured, throughout, to direct my attention to those objects which are of greatest importance to a Civil-Engineer. My observation embraced many of the principal Sea-ports, and navigable Rivers, two of the Great Lakes, the principal Canals, Railroads, Bridges, and other means of communication, and the most remarkable of the works for supplying the cities with water. The Steam-navigation of those countries, and the system of Lighthouses established along their coasts, also came incidentally under my notice, as well as some other points of more or less interest and importance.

I was well aware, before leaving this country, that a field so extensive and varied could not be fully examined in so limited a period; but this rapid tour, though it has not afforded that full measure of information upon many points of inquiry, which, had my time permitted, it would have been my endeavour to procure, has fully answered my purpose, by giving me a general view of the state of Civil-Engineering in America.

Having in the course of this journey seen a good deal that was entirely new to me, I have been induced to lay before my professional brethren the information thus obtained. It is true that Civil-Engineering, as practised in America, is not always applicable to the circumstances of Europe; but still the modifications to which it is subject in a new country may prove useful, by suggesting va-

CONTENTS.

CHAP. I.—Harbours.

CHAP. II.—Lake Navigation.

CHAP. III.—River Navigation

CHAP VI —CANALS.

CHAP. VII.—ROADS.

CHAP. VIII.—BRIDGES.

CHAP. XII

ERRATUM.

Page 200, line 12, *for* twenty-nine locks, *read* one hundred and twenty-nine locks,

SKETCH

OF

AMERICAN ENGINEERING.

CHAPTER I.

HARBOURS.

Natural facilities for the formation of Harbours on the American Coast—Tides—Construction of Quays, and Jetties—Cranes—Graving Docks—Screw Docks—Hydraulic Docks—Landing Slips, &c.—New York—Boston—Philadelphia—Baltimore—Charleston—New Orleans—Quebec—Montreal—Halifax.

THE eastern and southern coasts of North America are indented by numerous bays and sheltered sounds, which afford natural facilities for the formation of harbours more commodious than any which works of art alone, however costly, could possibly supply, and to an extent of which, perhaps, no other quarter of the globe can boast. The noble rivers with which this country abounds, and its inland lakes, which, for expanse, deserve the name of seas, are subjects of great interest to the general traveller; but to the civil-engineer, who is more alive to the importance of deep water and good shelter in the formation of harbours, and who, at every step in the exercise of his profession, feels the difficulty, and is made aware of the expense, which attend the attainment of these indispensable qualities by artificial means, the natural harbours of the continent of North America afford a most interesting and instructive subject of contemplation.

The original founders of the sea-port towns on this coast appear to have been very judicious in their selection of situations for forming their settlements. The towns, if not placed at the mouths of fine navigable rivers, in most cases possess the advantages of sheltered anchorages, with deep water, and accommodation for all classes of vessels. The chief object in founding most of the towns seems to have been the formation of a port for shipping, or the cultivation of a valuable adjacent tract of country watered by a navigable river; in which latter case the harbours do not always possess the same natural advantages, but stand in need of works for their improvement, which would involve a greater expenditure of capital, and occupy more time in their execution, than a country, as yet new in the arts, has been disposed to bestow upon them. Viewing the harbours of America generally, however, no one can fail to be struck with their importance, and, in connection with its inland navigation, convinced of their mighty effect in advancing the prosperity of that enterprising country.

The largest ports of North America are Quebec, Halifax, and Montreal, in the British dominions, and

Boston, New York, Philadelphia, Baltimore, Charleston, and New Orleans, in the United States. Besides these ports, there are many towns on the coast, of later origin, having less trade and importance, but nevertheless possessing splendid natural facilities for the formation of harbours.

I was fortunate enough to visit many of the American ports, and in most of them, I found that accommodation for vessels of great burden had been obtained in so satisfactory a manner, and at so small an expense, as could not fail to strike with astonishment all who have seen the enormously costly docks of London and Liverpool, and the stupendous asylum harbours of Plymouth, Kingstown, and Cherbourg. I have little hesitation in saying, that the smallest of the post-office packet stations in the Irish Sea has required a much larger expenditure of capital, than the Americans have invested in the formation of harbour accommodation for trading vessels along a line of coast of no less than 4000 miles, extending from the Gulf of St Lawrence to the Mississippi.

The American packet-ships trading between New York and the ports of London, Liverpool, and Havre, are generally allowed to be the finest class of merchant-vessels at present navigating the ocean; and for their accommodation we find in England the splendid docks of London and Liverpool, and in France the docks of Havre. An European naturally concludes that a berthage no less commodious and costly awaits their arrival in the ports to which they sail; but great will be his astonishment when, on reaching New York, the same fine vessel which lately graced the solid stone-docks of Europe, is moored by bow and stern to a wooden quay; and, on leaving the vessel, he will not fail to miss the shade of a covered verandah enclosed within high walls, the characteristic of a British dockyard, and will have any thing but pleasant sensations when he is ushered forth upon a hastily constructed wooden jetty, which, in certain states of the weather, is deeply covered with mud, and generally affords a footpath far from agreeable.

This state of things strikes a foreigner, on first landing in America, in a very forcible manner. The high, and in some cases superfluous, finish, which the Americans bestow on many of their vessels employed in trading with this country, lead those who do not know the contrary to expect a corresponding degree of comfort, and an equal display of workmanship, in the works of art connected with their ports; and it strikes one at first sight as a strange inconsistency, that all the works connected with the formation of the harbours in America should be of so rude and temporary a description, that, but for the sheltered situations in which they are placed, and other circumstances of a no less favourable nature, the structures would be unfit to serve the ends for which they were intended. But, when we come to inquire into the reasons for this difference between the construction of the European and

American harbours, they soon become apparent and satisfactory. The difficulties and expense encountered in the formation of most European harbours, have arisen chiefly from the necessity of constructing works of a sufficient strength to withstand the violence of a raging sea to which they are in general exposed, or in obtaining a sufficient depth of water, by the construction of docks or other means, to enable the vessels frequenting them to lie afloat at all times of tide. In Britain, these difficulties in a great measure arise from the narrowness of our country, which necessarily contains but a small extent of inland waters, whose quantity and currents, when compared with the bays and rivers on the American coast, are agents too unimportant and feeble to produce, without recourse to artificial means, the depth or shelter required in a good harbour. The Americans, on the contrary, among the numerous large bays and sounds by which their coasts are indented, have the choice of situations for their harbours, perfectly defended from the surge of the ocean, and requiring no works, like the breakwaters of Plymouth and Cherbourg, for their protection; and the basins formed and scoured by their large navigable rivers afford, without resorting to the construction of docks like those of Liverpool, London, Leith, or Dundee, natural havens, where their largest vessels lie afloat at all times of tide within a few paces of their warehouse doors.

The kind of workmanship which has been adopted

in the formation of the American harbours is almost the same in every situation; and the harbours generally bear a strong resemblance to each other in the arrangements of the quays, and even in their localities. This renders a detailed description of the works of more than one harbour unnecessary; and, for the purpose of giving an idea of an American harbour, I would select that of New York, because it undoubtedly ranks as the first port in America, and is, in fact, the second commercial city in the world, the aggregate tonnage of the vessels belonging to the port being exceeded only by that of London.

The island of Manhattan, in the state of New York, is about fifteen miles in length, and from one to three miles in breadth. The city of New York is situate on the southern extremity of this island, in north latitude 40° 42′, and west longitude 74° 2′ from Greenwich. It was founded by the Dutch in the year 1612, and it now contains a population of about 300,000 inhabitants, and measures about ten miles in circumference. On the east, the shore of Manhattan Island is washed by the sound which separates it from Long Island, and on the west by the estuary of the river Hudson, which, as far up as Albany, is more properly an arm of the sea than a river, the stream itself being small and contemptible. These waters, which have received from the Americans the appellation of the East and North Rivers, meet at the southern extremity of the island of Manhattan, and at their junction form the spacious bay and harbour of New York, the great emporium of the western hemisphere.

The Bay of New York, which extends about nine miles in length and five miles in breadth, has a communication with the Atlantic Ocean through a strait of about two miles in breadth, between Statten Island and Long Island. This strait is called "The Narrows;" and on either shore stands a fort for protecting the entrance to the harbour. This magnificent bay, which is completely sheltered from the stormy Atlantic by Long Island, forms a noble deep-water basin, and offers a spacious and safe anchorage for shipping to almost any extent, while the quays which encompass the town on its eastern, western, and southern sides, afford the necessary facilities for loading and discharging cargoes. The shipping in the harbour of New York, therefore, without the erection of breakwaters or covering-piers, is, in all states of the wind, protected from the roll of the Atlantic. Without the aid of docks, or even dredging, vessels of the largest class lie afloat during low water of spring-tides, moored to the quays which bound the seaward sides of the city; and, by the erection of wooden jetties, the inhabitants are enabled, at a very small expenditure, to enlarge the accommodation of their port, and adapt it to their increasing trade.

The situation of New York is peculiarly favourable for the extensive trade of which it has become the seat, by the nearness of its harbour to the ocean; the quays being only about eighteen miles from the shore of Sandy Hook, which is washed by the waters of the Atlantic. This naturally makes the communication more direct and easy, as a very short time elapses between making land and mooring at the quay ; and all the anxiety which is experienced after falling in with the European land, in a coasting navigation of several days, before the mariner terminates his cares by docking his vessel in Liverpool or London, or in any other port of Great Britain, is thus avoided. I may mention, in illustration, that I left the quays of New York at half-past eleven on the forenoon of the 8th of July 1837, in the "Francois Premier" packet-ship, Captain Pell, for Havre, with a very light breeze from the north-west ; and, at seven o'clock on the evening of the same day, our vessel was gliding through the Atlantic with nothing in sight but sky and water. This case is strongly contrasted with what took place on my outward passage, on which occasion I left Liverpool, under no less advantageous circumstances, on the 12th of March of the same year, in the "Sheffield" packet-ship, Captain Allen ; but we did not clear the Irish land till two days after our leaving port.

The perpendicular rise of tide in the harbour of New York is only about five feet. The tidal wave, however, increases in its progress northwards along the coast, till at length, in the Bay of Fundy, it attains the maximum height of 90 feet. Towards the south, on the contrary, its rise is very much decreased; and, in the Gulf of Mexico, is reduced to eighteen inches, while on the shores of some of the West India Islands it is quite imperceptible.

A bar extends from Sandy Hook to the shore of Long Island, across the entrance to the harbour. Over this there is a depth of twenty-one feet at low water, which is sufficient to float the largest class of merchant-vessels.

The wharfs erected for the accommodation of the shipping of New York are formed entirely of timber and earth, in a very rude and simple manner. A row of wooden piles, driven close to each other into the bed of the river, forms the face-work of the quay, which is projected from the shore as far as is necessary to obtain a depth of water sufficient to float the largest class of vessels at all times of the tide. The situation of New York, in this respect, is very favourable, as deep water is very generally obtained at the distance of from forty to fifty feet from the margin of the water. The piles, of which the face-work of the piers is composed, are driven perpendicularly into the ground, and are secured in their place by horizontal wale-pieces or stretchers, bolted on the face of the quay, and running throughout its whole extent. Diagonal braces are also bolted on the inside of the piles, and beams of wood are connected to the face-work, and extend behind it to the shore, in which they are firmly embedded. These beams act both as struts and ties, serving to counteract the tendency of lateral pressure, whether acting externally or internally, to derange the line of quay. The void between the perpendicular piles, which form the face-work and the sloping bank rising from the margin of the water, is generally filled up with earth, obtained in the operation of levelling sites and excavating foundations for the dwellings and warehouses of the city. This hearting of earth is carried to the height of about five feet above high water of spring-tides, at which level the heads of the piles, forming the face-work, are cut off, and the whole roadway or surface of the quay is then planked over. The planking used in forming the roadway of the quay is, in some cases, left quite exposed; but, in general, where there is a great thoroughfare, the surface of the quays is pitched with round water-worn stones, and corresponds, in appearance and level, with the adjacent streets. The following cross section of one of the wharfs, will shew more clearly the manner in which they are constructed.

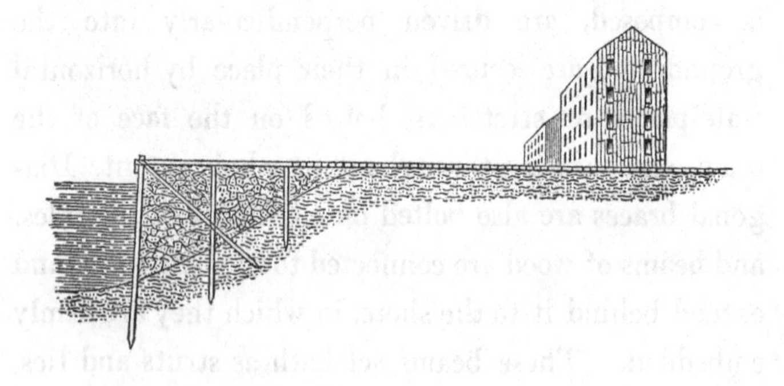

A continuous line of wooden quay-wall, constructed in this manner, surrounds the city of New York on its eastern, western, and southern sides; and the inhabitants are still rapidly extending their harbour accommodation to meet the wants of increasing trade, which has now become so great, that the wooden wharf-walls, by which the city is surrounded, have attained a length of no less than seven miles. Numerous jetties, of the same construction as the continuous quay-wall already described, project into the harbour from its face, at distances of from three to four hundred feet apart.—The jetties are generally from two to three hundred feet in length, and from fifty to sixty feet in breadth. The vessels frequenting the harbour, for the purpose of discharging or loading their cargoes, are moored in the bays formed between these projecting jetties, where they lie closely penned together, waiting their turn to get a berth alongside the wharfs.

The wood-work in the quays and jetties is of a very rude description. The timbers employed in their construction are seldom squared, and never, in any case, protected by paint or coal-tar from the destroying effects of the atmosphere. Wood is so plentiful in America, that to repair, or even construct works in which timber is the only material employed, is generally regarded as a very light matter.

The fixed crane for raising great weights, which is so generally used in the quays of Europe, is not employed in New York, nor, in fact, in any of the American ports. There, vessels generally discharge and take in cargo with a purchase hung from the yard-arm. Tackling, attached to moveable sheer-poles or derricks, is also in pretty general use in some of the ports; but this apparatus proves a very poor substitute for fixed quay-cranes, which are certainly of great convenience and utility in a shipping port.

The want of proper accommodation for vessels requiring repair is much felt by the shipping frequenting the American ports. The construction of an effective graving dock is, under any circumstances, an operation of considerable expense; but, in situations where the rise of tide is small, the difficulties encountered in its construction, and the inconvenience and expense attending the use of it when completed, prove a great bar to the introduction of this useful appendage to a dock-yard. It is, in a great measure, owing to these circumstances that graving docks, for the repair of trading vessels, are not used in the American ports; in the most important of which, the perpendicular rise of tide is so small, as to lessen, in a great degree, the advantages which, under more favourable circumstances, would be derived from their introduction.

The only graving docks at present existing in North America, are those which have been erected for the use of the Navy by the Government of the United States, in the Navy-yards of Boston in Massachusetts, and Norfolk in Virginia. These docks have been formed of such a size, as to admit, with ease, the largest class of government vessels belonging to the American Navy. The dock of Boston measures 341 feet in length, and 80 feet in breadth, and has a depth of water of 30 feet. But, although the depth of water in the dock is 30 feet at high water of spring tides, the fall of the tide is only 13 feet, which leaves 17 feet of water to be pumped out of the dock by means of a steam-engine every time a vessel is admitted for repair, an operation both tedious and expensive. The material used in their construction is a grey-coloured granite from Quincy in Massachusetts, and, as far as regards workmanship and general execution, they are inferior to no marine works which I have ever seen. These graving docks are believed to have cost about L. 152,000 each. They are the finest specimens of masonry which I met with in America, and are equally creditable to the government of the United States, and to Mr Baldwin, the engineer under whose direction they were constructed.

In the American harbours the method of careening or laying vessels on their sides to get at their lower timbers, is still often resorted to. I, however, met with three different mechanical arrangements for raising vessels from the water, when decay or damage renders this operation necessary for effecting their repair. In one of these arrangements, the requisite object is attained by the use of an inclined plane (on the well-known principle of Morton's patent-slip, but of a very rude description), on which vessels are drawn ashore by means of a system of wheel-work driven by a steam-engine.

The second method, which savours more of originality, is called the Screw-dock, the operation of which I witnessed on one occasion in the harbour of New York. The vessel to be raised by this apparatus was floated over a platform of wood, sunk to the depth of about ten feet below the surface of the water, and suspended from a strongly built wooden frame-work by sixteen iron screws four and a half inches in diameter. This platform has several *shores* on its surface, which were brought to bear equally on the vessel's bottom, to prevent her from canting over on being raised out of the water. About thirty men were employed in working this apparatus, who, by the combined power of the lever, wheel and pinion, and screw, succeeded, in the course of half an hour, in raising the platform, loaded with a vessel of 200 tons burden, to the surface of the water, where she remained high and dry, suspended between the wooden frames. At Baltimore, I saw a large screw dock, constructed on the same principles, on which the platform for supporting the vessel was suspended by forty screws of about five inches in diameter.

The last of those methods to which I have alluded, is an apparatus called the Hydraulic-dock, a beautiful application of the principle of Bramah's press, to produce a power capable of raising vessels of 800 tons burden. In this apparatus, as in the screw-dock, the vessel is raised on a platform swung between two frames. In the hydraulic dock, however, the platform is suspended by forty chains, twenty on each side, which

PLATE 1

Sketch shewing the principle of the Hydraulic Dock at New York.

Fig. 1.

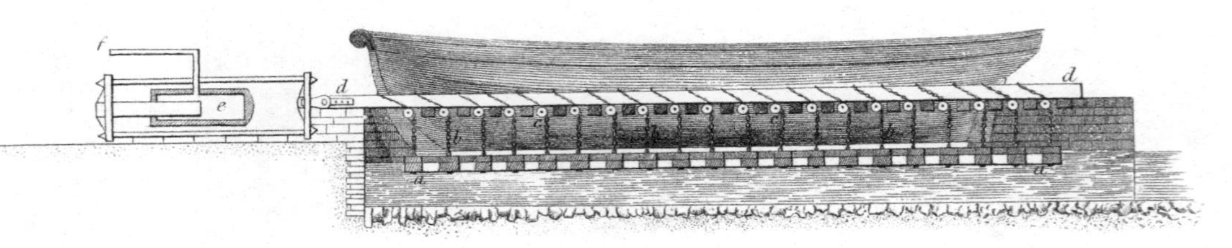

Fig. 2.

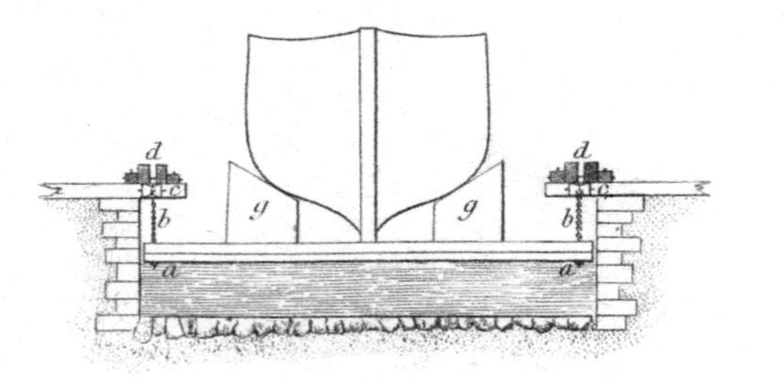

Stevenson's Sketch of the Civil Engineering of North America.

Thomas Stevenson, Delt.

Published by John Weale, 59, High Holborn. 1838.

Geo. Aikman, Sculpt.

pass over cast-iron pullies, supported on the top of the wooden frame-work. The lower ends of the chains are fixed to the platform, and the upper ends to a horizontal beam of wood, which is attached by means of a crosshead to the ram of a hydraulic engine. When the ram, therefore, which is placed in a horizontal position, is moved, by the injection of water into the cast-iron cylinder in which it works, the motion is communicated to the horizontal beam, and thence, by the suspending chains, to the platform bearing the vessel, which is thus slowly raised to the surface.

Plate I. is a sketch illustrative of the principles on which this apparatus is constructed. Fig. 1. is a longitudinal view, and Fig. 2. an end view of the platform and vessel. In both of these views, letters *a a a a* represent the platform; *b b b b b*, the suspending chains; *c c c c*, the pullies on which they run; *d d d d*, the horizontal beam to which the chains are attached; *e*, the hydraulic engine; and *f*, the injection-pipe by which the water is forced into the ram.

The cylinder and ram of the particular apparatus which I saw, were made in England, at the works of Messrs Bolton and Watt. The fixtures of the cylinder are embedded in a large mass of masonry, so as to render it quite immoveable. The perfect stability of this part of the apparatus is obviously of the highest importance, as the safety of the suspended vessel depends in a great measure on the attainment of this object. The external diameter of the water cylinder is twenty-eight inches, and its internal diameter is twelve inches. The ram which works in it is eleven inches in diameter, and ten feet in length. There are several racks attached to the apparatus, for supporting the platform, and taking part of the weight off the ram after the vessel is suspended. When she is ready to be lowered, these racks are unshipped, and the water being permitted to escape through a small aperture provided in the cylinder for that purpose, the vessel slowly descends into the water. The water is injected into the cylinder by a high-pressure steam-engine, of six horses' power, and the attendance of four persons is all that is necessary to raise a vessel of 800 tons register. The perpendicular *lift* of these docks is ten feet, which is found to be sufficient : the rise of tide in New York harbour being only five feet at spring tides, renders a greater height unnecessary.

The Screw and Hydraulic docks belong to a party of private individuals, called the "New York Screw-dock Company," who derive a considerable revenue from raising vessels by their ingenious apparatus. The following are their terms :—

For vessels under 75 tons, £3 per day.
Single-decked vessels of 75 tons and upwards, 10d. per ton per day.
Double-decked vessels of 75 tons and upwards, 1s. 0½d. per ton per day.

After the first day the charge is

For vessels under 170 tons, £3 per day.
For all vessels of 170 tons and upwards, 4½d. per ton per day.
Cargo or ballast is charged at the rate of 1s. 0½d. per ton.

The wharfs in the harbour of New York, are in general the property of private individuals, possessing the land on the margin of the river. Some of them also belong to the Corporation of New York. The wharfage dues are collected by the owners of the respective quays, and vary in their rates according to the local advantages which the sites possess, and the pleasure of the parties to whom they belong.

Vessels have, occasionally, been damaged while lying at the quays of New York, by the vast masses of floating ice which, upon the breaking up of the frost, are brought down from the interior of the country by the waters of the Hudson. For the protection of shipping against the recurrence of such accidents, which, however, are liable to affect only the vessels lying on the western side of the town, the erection of a breakwater in the river above New York harbour, has been for some time contemplated.

The trade of this great port is generally more or less interrupted by ice, for about a month every winter, and the river Hudson at New York has, once or twice, been covered by a coating of ice so thick as to afford a safe road for carriages. This, however, happens very rarely ; but such is the severity of the New York winter, that the omnibuses, and other wheel-carriages employed in running in the city, are always laid up for the space of five or six weeks during the depth of winter, and their places supplied by sledges, which run on the hardened snow.

The large suburb called Brooklyn occupies the shore of Long Island, directly opposite to New York. It is separated from the town by Long Island Sound, which at this point is about one-third of a mile in breadth, and forms part of the harbour of New York. One of the United States' navy-yards has been established at Brooklyn, which is also in other respects a place of considerable trade and importance. A constant communication is kept up between it and New York, by means of numerous steam-boats, which cross every five minutes, adding greatly to the bustle and confusion of this busy and crowded part of the harbour.

The stoppage and inconvenience which a bridge across the sound in this situation would occasion to the shipping, has prevented its erection, but the spirited inhabitants have had designs under their consideration for connecting the opposite shores by means of such a work, and also by the formation of a tunnel passing under the bed of the river, similar to that at present in progress under the Thames at London. The steam ferry-boats, however, are so well managed, that the want of a more constant means of communication is not much felt. They are twin boats with the paddle-wheel placed in the centre, and in their general construction resemble those at one time used on the ferries of the Tay at Dundee, and the Mersey at Liverpool.

The landing slips between which they ply are very

convenient and suitable for situations where the rise of tide is not great. The slip consists of a large platform of wood, the landward extremity of which is attached to the edge of the quay by moveable hinge-joints admitting of its free motion. The seaward extremity of this platform rests on a floating tank, and has the same elevation above the surface of the water as the deck of the ferry-boat. The outer extremity of the platform which rests on the floating tank, is thus elevated or depressed with the rise and fall of the tide, but always remains on a level with the steam-boat's deck, and affords during high-water a level road, and during low-water an inclined plane, for the passage of carriages and passengers between the vessel and the land.

Before quitting the subject of harbours, I shall make a few general remarks on some of the other American ports of consequence.

Boston, in Massachusetts, is generally supposed to rank next in importance to New York and New Orleans. The town is situated at the head of Massachusetts Bay, which extends over about fifty miles of coast between Cape Ann and Cape Cod, and contains within its limits many excellent anchorages. Boston Bay, in which the harbour has been formed, is a sheltered inlet of about seventy-five square miles in extent, enclosed by two necks of land, which so nearly approach each other as to leave only a very narrow entrance communicating directly with the Atlantic.

The exports from Boston are of a varied nature, consisting chiefly of the produce and manufactures of that part of the United States called New England. The population of the town is about 80,000. Its situation is curious. Placed on a peninsula having deep water close in-shore, and almost entirely surrounding it, it is connected with the adjoining country by means of a dam and seven wooden bridges, of which the most extensive is about a mile and a half in length. The dam consists of an embankment of earth 8000 feet in length, enclosed between two stone retaining-walls. It serves the double purpose of affording a means of communication, and also forming a large basin, in which the tide-water being collected, a water power is created for driving machinery.

The quays at Boston are constructed in the same style, and of the same materials, as those of New York, but more attention has been paid by the builders to the durability of the work. Some of the wharfs extend about a quarter of a mile into the harbour, and are of sufficient breadth to have a row of warehouses built on them. The rise of tide in Boston Harbour is thirteen feet in spring and nine feet in neap tides. In the suburb called Charlestown, which is connected with Boston by means of three wooden bridges, is situate the navy-yard of the United States, and the graving-dock already noticed.

Philadelphia is a town of 230,000 inhabitants, and stands on a peninsula between the rivers Delaware and Schuylkill in the State of Pennsylvania. Its harbour is at the head of the ship navigation of Delaware Bay, a vast arm of the sea, which is navigable for vessels of the largest class as far as Philadelphia, a distance of about a hundred miles from the Atlantic Ocean. In the bay of Delaware the tide has generally a rise of only three feet, but it is sometimes much increased by the state of the winds.

The town of Baltimore contains a population of about 80,000 inhabitants, and lies on the north side of the river Patapsco, about fourteen miles from its mouth. The basin forming the harbour is a splendid sheet of water, in which it is said 2000 vessels could ride at anchor with ease.

Chesapeake Bay, which receives the waters of the river Patapsco, on which Baltimore stands, is navigable for 200 miles from the ocean, and forms an outlet for the trade of the ports of Baltimore, Annapolis, Washington, Fredericksburg, Richmond, and Norfolk, and receives the waters of the Susquehanna, Patapsco, Potomac, and James rivers. The rise of tide at Baltimore is about five feet, but is much influenced by the state of the wind, which has a great effect upon the waters of Chesapeake Bay.

Charleston, in North Carolina, is a port of considerable size, built on a tongue of land formed by the rivers Ashley and Cooper. There is a bar at the entrance of the harbour with only twelve feet of water on it at low tides, but within the bar there is a good anchorage. The rise of tide in this harbour is about six feet.

As I had it not in my power to visit the Mississippi, I cannot speak of the port of New Orleans from personal knowledge; but as it is certainly the most important in the southern states, I felt unwilling to omit all mention of it in this sketch, and therefore applied to my friend Captain Basil Hall, who has kindly sent me the following notice on the subject.

"You are quite right," says Captain Hall, "to include New Orleans in your list of American harbours, for though it is not strictly a sea-port, it answers all the purposes of one in a remarkable degree. New Orleans lies at the distance of about a hundred miles from the Gulf of Mexico, and the ebb and flow of the tide do not reach so high as the city. The level of the river is, however, subject to fluctuations, in consequence of the changes in the supply of water from the upper countries through which it flows. It rises from January to May, remains full all June and a part of July, after which it begins to fall, and goes on decreasing in height till September and October, when it is lowest. The perpendicular difference in height of the surface of the Mississippi at New Orleans, is about thirteen or fourteen feet, and when at its lowest, it is nearly on a level with the sea at the mouth of the river, so that the flow is then scarcely perceptible.

"In former times, before steam-navigation was known, there was great delay, and considerable difficulty as well as danger, in getting from the sea to New Orleans, in consequence of the opposing stream, the numerous shoals, and the very tortuous nature of the course, which rendered it scarcely possible to sail up all the way with the same wind. To these annoyances may be added the very bad nature of the anchoring ground every where, and the difficulty as well as risk of lashing large vessels to the banks of such a river. All these things rendered New Orleans a harbour highly objectionable in a nautical point of view.

" Now, however, that steam has got command of " time and space," New Orleans may be considered an excellent sea-port, safe, and as easy of access as of egress. I need not mention that there are at all times any number of steam-tugs ready to take ships down the river, or to bring them up. When I was there in April 1827, eleven years ago, several steam-boats left the city every evening about sunset, each having in tow one or more vessels astern, besides one, two, or three lashed on each aside, so that the boat was often quite hid by the cluster round her. In this way they proceeded down, and at daylight came to the bar which lies across the mouth of the river opening into the Gulf of Mexico. On reaching the sea, or rather before they reached it, the steam-boats cast off their companions, and left them to be taken in charge by their respective pilots, unless in cases of calm or contrary wind, when, of course, they got a tow into the offing.

"The most important service of these steam-boats, however, is to tow ships up the river, for although it is always troublesome, and often very dangerous, to drop down with the current from New Orleans to the sea, it can be and is done, even without the help of steam. But to make way upwards against the Mississippi is a most heart-breaking work without such aid, and now-a-days the attempt would be considered absurd. Accordingly, the steam-vessels which have carried down the ships during the night, and have launched them in safety over the bar into the salt sea, look about them for others, which having made the land, are ready to enter the river. These they seize upon, and either take in tow, or lash alongside of them, and tow up to New Orleans. Of course they cannot, as in the downward case, carry along with them such a cluster as they brought down, nor is it likely that they will often be called upon to exert their strength so far, for the ships arrive off the entrance of the river by one or two at a time, and are not prepared, as within the port, to start in bodies at a given time.

"In this way, it may be fairly stated, that New Orleans, though a hundred miles from the sea, is virtually one of the best and most accessible ports in the Union. It may be added, that, as all the ships lie alongside of the levée or embankment which separates the river from the city, and which serves the purpose of a perfectly commodious wharf, and as the water is always smooth, nothing can be more easy and secure than the communication, both for loading and unloading goods. The ships lie alongside of each other in tiers, and I have seldom seen, in any country, such a forest of masts.

" Abreast of the upper part of the city may be seen, in like manner, numerous tiers of steam-boats of gigantic dimensions, just arrived from, or preparing to start for, the upper countries, through which the Mississippi and its innumerable tributaries pass. And farther up in this most extraordinary of harbours, lie crowds of huge rafts, or arks, as they are called,—rude vessels without masts, which have dropped down the river, and are loaded with that portion of the produce of the interior which will not bear the expense of steam-cariage.

" At every hour,—I had almost said at every minute of the day,—the magnificent steam-boats which convey passengers from New Orleans into the heart of the western country, fire off their signal guns, and dash away at a rate which makes me giddy even to think of.

" I must now conclude this brief notice by regretting, that the limitation in your time did not allow you to visit, and to describe in detail, this most remarkable of all the wonderful commercial phenomena,—as it may be called,—which the great western confederacy of states presents to the traveller, namely, a mighty city built in the midst of one of the most unhealthy swamps on earth, and a port, 100 miles from the sea, which rivals, in all essential respects, that of New York or London ; possessing, moreover, an uninterrupted and ready communication with the interior parts of a vast continent, to the distance of thousands upon thousands of miles, every where rife with civilization, though, but a few years ago, the whole was one vast wilderness, the exclusive abode either of alligators, wild beasts, or savages !"

These are the most considerable ports in the United States ; but, in addition, it may not be amiss shortly to notice the following bays and sounds, which deserve attention, as many of them afford good anchorage and sheltered lines of navigation.

Passamaquoddy Bay is situate at the boundary between the British dominions and the United States. It receives the waters of the river St Croix, the boundary line between the two countries. The tide in it rises twenty-five feet.

Penobscot Bay receives the waters of the Penobscot river, and has a rise of tide of ten feet.

Narragansett Bay is navigable for vessels drawing sixteen feet of water to the town of Providence, which is about thirty-five miles from the sea. The town of Newport in this bay, though a place of little importance, has one of the finest natural harbours in America.

Long Island Sound lies between the mainland and Long Island, and extends in a north-easterly direction from New York harbour. It affords a sheltered line of navigation of about a hundred and twenty miles in extent.

Albemarle and Pamlico Sounds, in North Carolina, are more remarkable for their curious geological formation than for any advantages held out by them for navigation, for which the difficulties of their entrance and shallow water, wholly unfit them. The narrow stripes of land, by which these sounds are separated from the Atlantic Ocean, stretch along the coast for a distance of about two hundred miles, and extend about forty miles south of Pamlico Sound. They are very little elevated above the level of the sea, and from their alluvial formation appear to have been gradually deposited by the Gulf Stream, which flows from the Gulf of Mexico, charged with the sediment and earthy matters borne down by the Mississippi and other streams which discharge themselves into the Gulf of Mexico.

Chatham, Appalachee, and Mobile Bays, in the Gulf of Mexico, are not reported as possessing, in any extraordinary degree, the qualifications of good havens, and, as already noticed, there is very little rise of tide on this coast. It may also be mentioned, that the hot and unhealthy climate of all the southern ports of the United States, from Charleston to New Orleans inclusive, as well as the nature of the slave population of the southern states, renders them very unsuitable for the growth of that hardy race of seamen, of which the northern ports of the country are the true and only nurseries.

The naval-yards belonging to the Government of the United States are established at Boston, Portsmouth in New Hampshire, New York, Philadelphia, Washington, Norfolk in Virginia, and Pensacola in the Gulf of Mexico; and those of them which I had an opportunity of visiting seemed to be very well regulated. Considering the natural advantages held out by that country, and the abundance of fine timber produced in it, it is not surprising that the Americans have bestowed so much attention upon naval affairs, or that their efforts should have been crowned by so great success in the improvement both of inland and maritime navigation. The genius of the people for naval affairs is doubtless the birthright of their British origin, and their patrimony has been improved by the energy which characterises all their efforts.

Quebec is the seat of government of Lower Canada, and, in a commercial point of view, is the first port in the British dominions in America. It is situate at the junction of the river St Charles with the St Lawrence; and, though distant fully 700 miles from the Atlantic Ocean, the spacious and beautiful Bay of Quebec, formed by the junction of the two rivers, affords a noble deep-water anchorage for vessels of all sizes, and almost in any numbers.

The bay measures about three miles and three quarters in length, and two miles in breadth, and the water in some parts of it is twenty-eight fathoms in depth. The population of the town is about 22,000, and its trade consists in the export of wood, potash, and furs, the produce of Upper and Lower Canada. The rise of tide at Quebec is twenty-three feet in spring-tides, and the quays and wharfs there, as well as in the harbours of the United States, are constructed entirely of wood.

The ferry-boats at Quebec, plying between the opposite sides of the river, which is about a quarter of a mile in breadth, are propelled by horses and oxen. These animals are secured in small houses on the decks of the vessels ; and the effort they make in the act of walking on the circumference of a large horizontal wheel, produces a power which is applied to drive the paddle-wheels of the ferry-boat, in the same manner as the motion of the wheel in the tread-mill is applied to the performance of different descriptions of work. I have seen horse ferry-boats in Holland, and, I believe, they have also been used in America, in which the power was more advantageously applied by means of an apparatus like the gin of a thrashing-mill, in which case the horses are not stationary, but are made to walk in a circle, and the motion communicated by them to an upright shaft, is conveyed, by means of wheel-work, to the paddle-wheels of the vessel. A boat of this kind was used for some time in England, between Norwich and Yarmouth.

Montreal, which is 180 miles to the westward of Quebec, and 880 miles from the ocean, is at the head of the ship navigation of the St Lawrence, and considerably above the influence of the tide. The town is built on the island whose name it bears, which is situate at the junction of the Ottowa, or Grand River, with the St Lawrence. The quays and landing slips at Montreal are built of stone; and in this respect it differs from the other American ports which I have noticed. The material used in their construction is a blue limestone, which is very abundant throughout the greater part of Canada, and is much used in all building operations. The trade of Montreal is of the same description as that of Quebec, though not so extensive.

Halifax harbour is considered one of the finest in the world, and is calculated to afford anchorage for upwards of a thousand vessels of the largest class. It is a place of very considerable importance; for through it comes much of the trade of Nova Scotia; and it is the British post packet-station for Canada.

Such is a brief sketch of the construction and capabilities of some of the principal harbours of America, in the formation of which nature has done so much, that little has been left for the labour of man, and works of an extensive and massive description, and operations such as are found to be indispensable in rendering European harbours accessible or commodious, have there been found to be unnecessary. By erections of a temporary description, constructed of the wood produced in the operation of clearing their lands, the inhabitants have been enabled, along the whole line of coast, to afford, at a very small cost, accommodation for an extent and class of shipping, to obtain which, in any other quarter of the globe, would have involved an enormous investment of capital, and a much greater consumption of time.

CHAPTER II.

LAKE NAVIGATION.

Great Western Lakes—Ontario—Erie—Huron—Michigan—Superior—Welland Canal—Lake Harbours—Construction of Piers, Breakwaters, &c.—Buffalo—Erie—Oswego—Toronto—Kingston—Vessels employed in Lake Navigation—Violent Effects of Storms on the Lakes—Ice on the Lakes—Effects of Ice on the Climate—Lake Champlain.

THE great chain of inland lakes, whose vast expanse justly entitles them to the name of seas, are the largest bodies of fresh water in the known world, and constitute an important feature in the physical geography of North America. When viewed in connection with the River and Gulf of St Lawrence, by which their surplus waters are discharged into the Atlantic Ocean, ideas of magnitude and wonder are excited in the mind, which it is impossible to describe. But the effects which they produce on the commercial and domestic economy of the country are considerations far more important and striking. With the aid of some short lines of canal, formed to overcome the natural obstacles presented to navigation by the Falls of Niagara and the Rapids of the St Lawrence, these great lakes are converted into a continuous line of water-communication, penetrating upwards of 2000 miles into the remote regions of North America, and affording an outlet for the produce of a large portion of that continent, which, but for these valuable provisions of nature, must, in all probability, have remained for ever inaccessible.

The great western lakes of America are five in number :—Ontario, Erie, Huron, Michigan, and Superior. The extent of these lakes has been variously stated, and the several accounts which have been given of them, differ very considerably ; but the dimensions which I shall quote are taken partly from the work of Mr Bouchette, the Surveyor-General of Canada, and partly from the charts constructed by Captain Bayfield, of the Royal Navy.

Lake Ontario, the most eastern of the chain, lies nearest to the Atlantic. The River St Lawrence, which has a course of about a thousand miles before reaching the ocean, is its outlet, and flows from its eastern extremity. This lake is 172 statute miles in length, 59¼ miles in extreme breadth, and about 483 miles in circumference. It is navigable throughout its whole extent for vessels of the largest size. Its surface is elevated 220 feet above the medium level of the sea ; and it is said to be, in some places, upwards of 600 feet in depth. The trade of Lake Ontario, from the great extent of inhabited country surrounding it, is very considerable, and is, at this moment, rapidly increasing.

Many sailing vessels and splendid steamers are now employed in navigating its waters. Owing to its great depth, it never freezes, except at the sides, where the water is shallow ; so that its navigation is not so effectually interrupted as that of the comparatively shallow Lake Erie.

The most important places on the Canadian or British side of Lake Ontario, are the city of Toronto, which is the capital of Upper Canada, and the towns of Kingston and Niagara, and, on the American shore, the towns of Oswego, Genesee and Sackett's Harbour. Lake Ontario has a direct communication with the Atlantic Ocean, in a northerly direction, by the St Lawrence, and in a southerly direction by the river Hudson and the Erie Canal, with which it is connected by a branch canal, leading from Oswego to a small town on the line of the Erie Canal called Syracuse.

Lake Erie is about 265 miles in length, from thirty to sixty miles in breadth, and about 529 miles in circumference. The greatest depth which has been obtained in sounding this lake is 270 feet, and its surface is elevated 565 feet above the level of the Hudson at Albany. Its bottom is composed chiefly of rock. Lake Erie is said to be the only one of the chain in which there is any perceptible current, a circumstance which may, perhaps, be occasioned by its smaller depth of water. This current, which runs always in the same direction, and the prevailing westerly winds, are rather against its navigation. The shallowness of the water also, which varies from 100 to 270 feet in depth, renders it more easily and more permanently affected by frost, its navigation being generally obstructed by ice for some weeks every spring, after that of all the other lakes is open and unimpeded.

The principal towns on Lake Erie are Buffalo, Dunkirk, Ashtabula, Erie, Cleveland, Sandusky, Portland, and Detroit. Between forty and fifty splendid steam-boats, and many sailing-vessels, are employed in its trade, which is very extensive; and several harbours with stone-piers have been erected on its shores for their accommodation.

The surface of Lake Erie is elevated 322 feet above Lake Ontario, into which its water is discharged by the river Niagara. In the course of this river, which is only thirty-seven miles in length, the accumulated surplus waters of the four upper lakes descend over a perpendicular precipice of 152 feet in height, and form the "Falls of Niagara." These falls, with the rapids which extend for some distance both above and below them, render seven miles of the river's course unfit for navigation. The unfavourable structure of the bed of the river Niagara,—the connecting link between Lakes Erie and Ontario,—for the purposes of navigation, induced a company of private individuals, assisted by the British Government, to construct the Welland Canal, by which a free passage from the one lake to the other is now afforded for vessels of 125 tons burden.

This undertaking was commenced in the year 1824,

and completed in 1829, five years having been occupied in its execution. The expense of the works connected with it is said to have been about L.270,000.

The canal extends from Port Maitland on Lake Erie to a place called Twelve-Mile Creek on Lake Ontario. Its length is about forty-two miles ; its breadth at the surface of the water is fifty-six feet, and at the bottom twenty-six feet, and the depth of water is eight feet six inches. The whole perpendicular rise and fall from the surface of Lake Ontario to the summit level, and thence to Lake Erie, is 334 feet, which is overcome by means of thirty-seven locks of various lifts, measuring one hundred feet in length and twenty-two feet in breadth, most of which are formed of wood. The most considerable work occurring on the Welland Canal is an extensive excavation of forty-five feet in depth, from which 1,477,700 cubic yards of earth, and 1,890,000 cubic yards of rock, are said to have been removed.

Lake Erie is connected by the Erie Canal with the river Hudson and the Atlantic Ocean, and again by the Ohio Canal with the river Ohio and the Gulf of Mexico. The Erie Canal is 363, and the Ohio Canal 334, miles in length. I shall advert more particularly, however, to the construction and details of the canal works in North America in another section.

Lake Huron is about 240 miles in length, from 186 to 220 miles in breadth, and 1000 miles in circumference. The outline of this lake is very irregular, and Mr Bouchette says of its shores, that they consist of "clay cliffs, rolled stones, abrupt rocks, and wooded steeps." Its connection with Lake Erie is formed by the river St Clair, which conveys its water over a space of thirty-five miles into a small lake of the same name, of a circular form, and about thirty miles in diameter, from whence the river Detroit, having a course of twenty-nine miles, flows into Lake Erie. The communication between the two lakes is navigable for vessels of all sizes.

Lake Michigan is connected with Lake Huron by the navigable strait Michillimackinac, in which is situate the island of Mackinaw, now the seat of a custom-house establishment, and a place of considerable trade. Lake Michigan is about 300 miles in length, seventy-five miles in breadth, and 920 miles in circumference, having a superficies of 16,200 square miles. It is navigated by many steamers throughout its whole extent. The principal towns on the lake, the southern shore of which has now become the seat of many prosperous settlements, are Michigan, Chicago, and Milwawkie. The Illinois river takes its rise near the shores of Lake Michigan, and flows into the Mississippi; and a canal, for the purpose of connecting their waters, is now in progress; an improvement which, when completed, will form a second water-communication, extending from the Gulf of St Lawrence to the Gulf of Mexico, a distance of upwards of 3000 miles,—the other communication being that already alluded to between Lake Erie and the Ohio by a canal from Cleveland to Portsmouth.

Lake Superior is connected with Lake Huron by the river St Mary. This river, which is about forty miles in length, has a fall of twenty-three feet on the whole length of its course, and is navigable only for small boats. As yet the march of improvement has not penetrated to this remote region, but ere long Lakes Superior and Huron, like Erie and Ontario, will probably be connected by a canal. Lake Superior is about 360 miles in length, 140 miles in breadth, and 1116 miles in circumference; the depth is in some places said to be 1200 feet, and its surface is 627 feet above the level of the sea. Its bottom consists of clay and small shells. This lake is the largest body of fresh water known to exist; and although surrounded by a comparatively desert and uncultivated country, at a distance of nearly 2000 miles from the ocean, and at an elevation of 627 feet above its surface, it is navigated by steam-boats and sailing vessels of great burden, which are reported to be not inferior to the craft navigating the lower lakes.

From what has been said regarding the great western lakes, it will easily be believed that, notwithstanding the secluded situation which they hold in the centre of North America, far removed from the ocean and from intercourse with the world at large, their waters are no longer the undisturbed haunt of the eagle, nor their coasts the dwelling of the Indian. Civilization and British habits have extended their influence even to that remote region, and their shores can now boast of numerous settlements, inhabited by a busy population, actively engaged in commercial pursuits. The white sails of fleets of vessels, and the smoking chimneys of numerous steamers, now thickly stud their wide expanse, and beacon-lights, illuminating their rocky shores with their cheering rays, guide the benighted navigator on his course. Every idea connected with a *fresh-water lake*, must be laid aside in considering the different subjects connected with these vast inland sheets of water, which, in fact, in their general appearance, and in the phenomena which influence their navigation, bear a much closer resemblance to the ocean than the sheltered bays and sounds in which the harbours of the eastern coast of North America are situated, although these estuaries have a direct and short communication with the Atlantic Ocean.

The whole line of coast formed by the margins of the several lakes above enumerated, extends to upwards of 4000 statute miles. There are several islands in Lake Superior, and also at the northern end of Lake Michigan, but the others are, generally speaking, free from obstructions. They have all, however, deep water throughout their whole extent, and present every facility for the purposes of navigation.

It was not till the year 1818, that the navigation of the lakes had become so extensive and assumed so important a character, as to render the erection of lighthouses necessary and expedient, for insuring the safety of the numerous shipping employed on them. Since that period, the lighthouses have been gradually increasing, and, on the American side of the lakes, they now amount to about twenty-five in number, besides about thirty beacons and buoys, which have been found of the greatest service.

About the same period at which the introduction of lighthouses was considered necessary, some attention was also bestowed on the subject of lake harbours. Many which formerly existed, were then improved and enlarged, and others were projected, and the works connected with them are now either finished, or are drawing to a close. I visited several of these ports on Lakes Erie and Ontario, which have good sheltered anchorages, with a sufficient depth of water at their entrances for the class of vessels frequenting them. But good harbour accommodation is by no means so easily obtained on the shores of the lakes, as, generally speaking, on the sea coast of the United States. Most of the lake harbours are formed in exposed situations, and as regards the expense and durability of the several works executed in their formation, are much better calculated to resist the fury of the winds and waves, than the wooden wharfs of the sea-ports on the eastern coast of the country of which I have given a description. In connection with what has already been said on the subject of the harbours of the American coast, I shall give a brief sketch of some of those which came immediately under my notice on the shores of the lakes.

The town of Buffalo stands at the eastern corner of Lake Erie in the state of New York, and contains a population of about 16,000. As regards the number of its inhabitants and the extent of its commercial transactions, it is the most important place on the lakes, being in fact the New York of the western regions. From the month of June till the month of December inclusive, during which period the navigation of the lakes is generally open and unimpeded by ice, between forty and fifty steam-boats, varying from 200 to 700 tons register, are constantly plying between Buffalo and the several ports on the shores of the lakes. Some of these steamers make regular voyages once a month to Chicago in Lake Michigan, a distance of no less than 965 miles; and one leaves the harbour of Buffalo twice every day, during summer, for Detroit, a distance of 325 miles. The New York and Erie Canal, the earliest, and perhaps the most important public work executed in the United States, which enters the lakes at Buffalo, has a great effect in increasing its trade and importance.

Buffalo is built at the mouth of a creek communicating with the lake, in which the harbour is formed. The wharfs in the interior of the harbour are made of wood, but the covering pier, and other works exposed to the wash of the lakes, are built of stone, and cost about L.40,000. The depth of water in the harbour is nine feet when the lake is in its lowest or summer water state. The following diagram represents a cross section of the covering pier, which has been erected for the purpose of protecting the shipping and tranquillizing the water within the harbour during heavy gales. It measures 1452 feet in length, and its form and construction are so very substantial, that one may fancy himself in some sea-port, forgetting altogether that he is on the margin of a fresh-water lake, at an elevation of more than 300 feet above the level of the ocean.

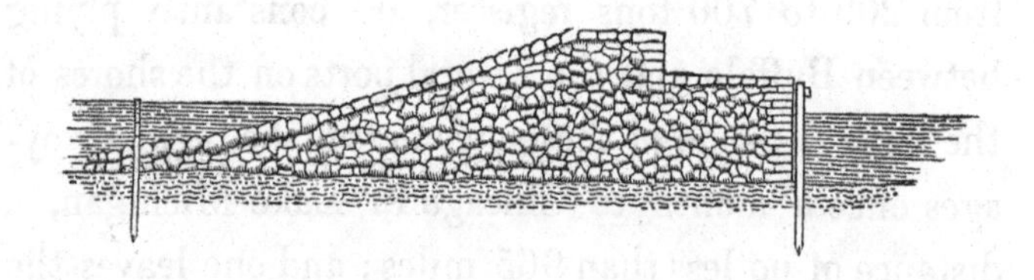

The top of the pier on which the roadway is formed, measures eighteen feet in breadth, and is elevated about five feet above the level of the water in the harbour. On the side of the roadway which is exposed to the lake, a parapet-wall five feet in height extends along the whole length of the pier, from the top of which, a talus wall, battering at the rate of one perpendicular to three horizontal, slopes toward the lake. This sloping wall is formed of a description of masonry, which is technically termed coursed pitching. Its foundations are secured by a double row of strong sheeting piles driven into the bed of the lake, and a mass of rubble *pierres perdues*, resting on the toe of the slope. The inner side of the pier, as shewn in the diagram, presents a perpendicular face toward the harbour, and is sheathed with a row of sheeting piles, driven at intervals of about five feet apart from centre to centre, to prevent the quay-wall from being damaged by vessels coming alongside of it.

The entrance to the harbour is marked by a double light, exhibited from two towers of good masonry built on the pier.

The workmanship and materials employed in erecting many of the other lake harbours, are of a much less substantial description than that adopted at Buffalo. The breakwater for the protection of Dunkirk Harbour on Lake Erie, for example, was formed in a most ingenious manner, by sinking a strong wooden frame-work filled with stones. The frame or crib was erected during winter on the ice over the site which it was intended to occupy. The ice was then broken, and the crib being filled with small stones, sunk to its resting place in the bottom of the lake.

Presque-Isle Bay, in which the town of Erie stands, is formed by the peninsula of Presque-Isle, on the shore of Lake Erie. This bay measures about one mile in breadth, and three miles in length, and affords a splendid anchorage for vessels of the largest size. It opens toward the north-west, and is sheltered from the waves of the lakes by two covering breakwaters, measuring respectively 3000 and 4000 feet in length, projecting from the shore, and leaving a space between their outer extremities of 300 feet in breadth, for the ingress and egress of vessels. Some other works of considerable extent are contemplated, to render this harbour still more safe and convenient.

Oswego, situate at the mouth of the Seneca River, on the southern shore of Lake Ontario, is a town of 6500 inhabitants, having a good harbour. It stands at the commencement of the branch canal, which connects the great New York and Erie Canal with Lake Ontario, and is the seat of several manufactories and mills driven by the Seneca River, on which there are some very valuable falls. The pier, which has been built at this place for the protection of the harbour, is a very good specimen of masonry, finished somewhat in the same style as that at Buffalo, and cost about L.20,000. The depth of water in the harbour is twenty feet, and it has a good harbour-light placed in a substantial tower of masonry at the extremity of the pier.

The works required in the construction of Buffalo, Erie, and Oswego harbours were done at the expense, and under the direction, of the government of the United States, who have also executed harbour-works of great extent, varying according to the nature of their situations, at the towns of Chicago, Michigan, Milwawkie and Green Bay in Lake Michigan; Detroit, Sandusky, Ashtabula, Portland, and Dunkirk, on

Lake Erie; and at Genesee and Sackett's Harbour on Lake Ontario. Sackett's Harbour is remarkable as having been the United States Navy-yard during the war.

The harbours on the Canadian or British shores of the lakes are, as yet, not so numerous. The principal ones are those of Toronto, Port Dalhousie, Burlington, Hungry Bay, and Kingston, on Lake Ontario; and Amherstburgh, and Put-in Bay on Lake Erie.

Toronto, the capital of Upper Canada, lies in a bay which is nearly circular, and measures about a mile and a half in diameter. It is sheltered from the lake by a projecting neck cf land called Gibraltar Point, on which the harbour-light is erected. This bay has a considerable depth of water, and affords an extensive and safe anchorage. Port Dalhousie is at the entrance of the Welland Canal, and has two piers, measuring respectively 200 and 250 feet in length, and also some pretty extensive works, connected with a basin for receiving timber. Kingston, situate at the eastern end of Lake Ontario, just at the point where the river St Lawrence flows out of the lakes, is the British Government Naval Yard. Navy Bay, in which it stands, is a good anchorage for vessels drawing eighteen feet of water, but is exposed to south and south-west winds. The British Government have also executed works in some of the other harbours on the Canadian side of the lakes.

The tonnage of most of the craft employed in the lake navigation is regulated by the size of the canals which have been constructed for the purpose of connecting the lakes, and facilitating the navigation of the St Lawrence. The locks of these canals are formed of such dimensions as to admit vessels of 125 tons burden, and consequently the lake craft, with a few exceptions, do not exceed this size. The steam-boats, however, and all the vessels which are employed exclusively in the navigation of one lake, are never required to enter the canals, and many of these are of great size; some of the new steamers being no less than 700 tons burden. The art of ship-building, which is practised to a considerable extent in almost every port, is greatly facilitated by the abundance of fine timber produced in the neighbourhood of the lakes; and to so great an extent has the art been carried on, that during the wars a vessel called the St Lawrence, of 102 guns, was launched by the British at Kingston, and another by the Americans at Sackett's Harbour, which measured 210 feet in length on her lower gun-deck

The vessels used in the lake navigation, and more especially the steam-boats, which I had frequent opportunities of examining, possess, in a much greater degree, the character of *sea-boats*, than the same class of vessels employed in the sounds and bays on the shores of the Atlantic; and the substantial masonry of which the piers and breakwaters on the lakes are composed renders these works also, as before noticed, much more capable of resisting the fury of the winds and waves than the wooden wharfs of the eastern coast of the country. The strength and durability of material which both the piers and the vessels present, are, at first sight, apt to appear superfluous in works connected with lake navigation. I was certainly impressed with this conviction when I first saw the stone-piers of Buffalo, which I have already described; and the sight of the steamer "James Madison," a strongly built vessel of 700 tons burden, drawing about ten feet of water, which plies between Buffalo on Lake Erie and Chicago on Lake Michigan, was in no way calculated to lessen the impression which the harbour had left; an impression which was heightened by the circumstance of my having, a short time before, examined the harbours on the eastern coast, and seen many of the slender fabrics, drawing from three to five feet of water, which navigate the bays and sounds in that part of the country. But, on inquiring more particularly into this subject, I was informed that these lakes are often visited by severe gales of wind, which greatly disturb the surface of their waters, and give rise to phenomena which one hardly expects to find in a fresh-water lake. In the opinion of many of the captains of the steamers with whom I conversed on this subject, the undulations created during some of these gales are no less formidable enemies to navigation than the waves of the ocean, so that the greatest strength in the hydraulic works and naval architecture of the lakes is absolutely necessary to insure their stability. I had not an opportunity, while in America, of witnessing the effects produced on the lakes by a gale of wind; but in many situations where their shores were exposed to a great expanse of water, and consequently with an in-shore wind to the action of waves having a long *fetch* and ample scope to develope themselves, I found many interesting indications of their occasional violence when under the action of a hurricane. In the harbour of Buffalo, for example, which is situated in the north-east corner of Lake Erie, and has an unobstructed expanse of water extending before it for a distance of about 180 miles, the effects of the waves are very remarkable. The pier at this place is built of blue limestone. The materials are small, and no mortar is used in its construction; but the stones are hammer-dressed, well jointed, and carefully assembled in the walls, and the structure, as before noticed, both as regards the materials of which it is built, and its general design, is calculated to stand a good deal of fatigue. On examining this pier, however, I was a good deal surprised to find that it was in some places very much shaken, and, more particularly, that several stones in different parts of the work had actually been raised from their beds; and I was told that this work, as well as most of the harbours on the lakes, has annually to undergo some repair of damage occasioned by the violence of the waves. I measured several of the stones which had been moved, and one of the largest of them, weighing upwards of half a ton, had been completely turned over, and lay with its bed or lower side uppermost.

I met with another striking example of the violence of the lake-waves on the road leading from Cattaraugus to Buffalo, which winds along the side of Lake Erie, in some places close to the water, and in others removed several hundred feet from its margin. The surface of this road is elevated several feet above the level of the lake ; but, notwithstanding this, many of the fine large trees, with which the whole country is thickly covered, have been rooted up and drifted across the road by the violence of the wind and waves, and now lie along its whole line piled up in the adjoining fields. Every winter's storm adds to these heaps of drifted timber, and they will doubtless continue to be enlarged till the increasing value of the lands on the margin of the lake, which, in their present state, are wholly useless in an agricultural point of view, renders the erection of works for their protection a matter of pecuniary interest to the proprietors.

The following extract also, from the Annual Report of the Board of the New York State Canals for 1835, shews the severity of the lake storms :—" The method of towing barges by means of steam-boats has been very successfully practised on the Hudson river ; but on the lakes, though a great many steam-boats have been in use for several years, the plan has not been adopted, because the steam-boats cannot manage barges in a storm. We have been informed of a proposition made to the proprietors of a steam-boat to take some canal boats from Buffalo to Cleveland; and it was accepted *only* on the condition, that, in the event of a storm, they should be at liberty to cut them loose at the risk of the owners.

"An intelligent gentleman, of several years' experience in navigating steam-boats, and the two last seasons on Lake Ontario, informs us, that he considered it impracticable, as a regular business, for steam-boats on the lakes to tow vessels with safety, unless the vessels were fitted with masts and rigging, and sufficiently manned, so as to be conducted by sails in a storm; that storms often rise very suddenly on these lakes, and with such violence as would compel a steam-boat to cut loose vessels in tow in order to sustain herself."

The most striking indications of the extreme violence of these storms are found in those parts of the coast where the lake is of great breadth, and where there is deep water close in-shore. On the other hand, in situations where the shores are contracted, or defended by islands, or where the lake is for some distance very shallow, the water does not appear to be so much agitated by the wind. Such facts regarding the lake-storms serve to indicate that the formation of those undulations in the sea, which prove so destructive to our marine-works, depends on the action of the wind, and is not necessarily connected with the great tidal wave occasioned by the attraction of the moon and sun, whose influence in affecting the level of the lakes is quite imperceptible, owing to the smallness of their area compared with that of the ocean. It also appears, from what has been stated, that, to the production of considerable undulations, capable of injuring marine-works, or endangering their stability, three conditions are necessary. *First,* That the sheet of water acted upon by the wind shall have a considerable area. *Second,* That its configuration shall be such, that the wind, moving over it in any direction, shall act upon its surface extensively, both in the directions of length and breadth. And, *third,* That the depth of the water shall be considerable, and unobstructed by shoals, so as to permit the undulations to develope themselves to a great extent, without being checked by the retardations caused by shallow water and an unequal bottom.

From my own observations, and from what I have heard regarding the form assumed by the lake-waves, and the effects produced by them, I am inclined to believe that they bear a strong resemblance to the undulations experienced, during gales of wind, in such land-locked bodies of water as the Irish Sea, which, it is well known, are very different from the long swell met with in the ocean. In all land-locked bodies of water, the waves are short and sudden in their movements, proving very destructive to whatever obstacle is opposed to their fury; but there is a characteristic slowness in the long movement of the ocean's swell, which, it is generally acknowledged, renders it less destructive to the marine-works exposed to its action than the waves produced in land-locked seas. It is confidently hoped that the experiments which Mr Russell and others are at present conducting, at the suggestion of the British Association, on the laws which regulate the undulation of fluids, may lead to some satisfactory results on this subject, so interesting in a speculative point of view, and so important to the engineer.

The great area presented by the surface of the lakes prevents any material variation in their level from taking place, which, in small bodies of water, would be the necessary consequence of the torrents annually poured into them from the melting snow. It is stated that a periodical rise of about two feet on the level of the lakes occurs every seven years; but the facts connected with this singular phenomenon do not appear to be very satisfactorily established. The water of the lakes and the river St Lawrence is remarkably pure and clear. Mr M'Taggart mentions, in his work on Canada, that a white object, measuring a foot square, may be seen at the depth of forty feet below the surface. From my own observation, however, I cannot say that the American lakes are, in this respect, more remarkable than the Lake of Geneva, the waters of which are certainly very transparent.

The rigour of a Canadian winter, covering the face of the country with snow, and congealing every river, lake, and harbour, produces a stagnation in trade, which cannot fail to have a bad effect on the commerce of the country and the habits of the people, who are compelled to complete their whole business transactions during the summer and autumn months, and remain in a state of comparative indolence during the remainder of the year. When this inauspicious and unfavourable state of things is kept in view, it is astonishing, and in the highest degree creditable, particularly to the inhabitants of the British colonies, that they, situated as they are on the least favourable side of the lakes, as far as climate is concerned, have made such rapid advances in agriculture and public works. Considering the lakes in a commercial point of view, it is impossible not to regret that their navigation is open for so very limited a period. For the space of, at least, five months in the year, the greater part of their surface is covered with a thick coating of ice; and the same sheet of water which, in summer, floats the vessel of 700 tons, and devastates the shores with its waves, becomes, in winter, a *highway* for the Canadian sledge. The centre of the lakes, where the water attains a considerable depth, is not frozen every season; but a vast sheet of ice is annually formed round their margins, which almost effectually puts a stop to navigation. Mr M'Taggart mentions that, in the year 1826, the ice at the margin of Lake Ontario was within half an inch of being two feet in thickness; and that, during the winter of the same year, Lake Chaudière was covered with a coating which measured no less than three feet six inches in thickness. He also made several experiments to ascertain the densities of lake and river ice, from which it appeared that the volumes of six cubic feet of lake, and eight cubic feet of river ice, were each equal when melted to five cubic feet of water. The ice on the rivers and lakes does not long retain a level surface. Large flaws make their appearance soon after it is formed, and the whole sheet gradually splits into pieces, which, being united together in great masses or hummocks, resist the action of the sun long after the disappearance of frost.

The period at which the lake navigation closes, is generally about the end of November or beginning of December, and this interruption is never removed before the first week of May. In 1837, the year in which I visited America, the navigation was not wholly open till the last week of May. On the 20th of that month, I passed down Lake Erie, on my way to Buffalo, in the steam-boat "Sandusky," on which occasion, even at that late period in summer, we encountered a large field of floating ice, extending as far as the eye could reach. Our vessel entered the ice about seven o'clock in the morning, and at twelve in the forenoon she had got nearly half way through this obstacle, when a breeze of wind sprung up, which, from its direction, had the effect of consolidating the field into a mass so compact, that our vessel being no longer able to penetrate it, was detained a prisoner, at the distance of about ten miles from Buffalo, the port for which she was bound. During the two following days, the efforts of our crew to free the vessel were unavailing, and so thick was the field of ice by which we were surrounded, that several of our less patient and perhaps more adventurous fellow passengers, made many fruitless attempts to reach the shore, which was only two or three miles distant, by walking over its surface. On the morning of the 23d, a breeze of wind fortunately loosened the ice, and our captain, after having seriously damaged his vessel in attempting to extricate her, succeeded in making his escape, and landed his unfortunate passengers during a torrent of rain, on the shores of the lake, far from any house, and ten miles from Buffalo, the place of our destination. The circumstance of there being upwards of two hundred passengers on board, and a great scarcity of provisions, together with the coldness of the weather, rendered our situation during the forty-eight hours of our imprisonment far from agreeable.

The country through which I travelled for some days before reaching the shores of the lakes, on my way from the Ohio River to Lake Erie, and also that part of it through which I passed on my route from the lakes to Quebec, presented all the indications of summer, every tree and shrub being in full foliage. In the immediate neighbourhood of Lake Erie, however, no signs of the approach of spring or returning vegetation were visible, though it was towards the end of May. The country surrounding the margin of the lake was bleak, and the trees were leafless, while the atmosphere was exceedingly damp, and the temperature indicated by the thermometer ranged from 32° to 35° of Fahrenheit. Such was the effect produced on the climate by this huge cake of floating ice, that it was almost impossible, from the state of the lake atmosphere, and the appearance of the surrounding country, to divest one's self of the idea that winter was not yet gone, although in fact the first month in summer was drawing to a close. This circumstance affords a striking example of the degree in which climate may be influenced by local circumstances; for, while the shores of Lake Erie presented this sterile appearance, and were still plunged in the depths of winter, the country in the neighbourhood of Quebec, although lying three degrees further north, was richly clothed with vegetation.

The transition from winter to summer in the northern parts of North America, is very sudden. There is no season in that country corresponding to our spring. The vast heaps of hardened snow and ice which have accumulated during the winter, remain on the ground long after the sun has attained a scorching heat, but it is not until his rays have melted and removed them, that the climate becomes really warm, and then the foliage being no longer checked by the cold produced by these masses of snow and ice, instantly bursts forth, and at that particular time a single day makes a marked difference on the face of the country.

The only other body of fresh water in North America demanding attention, is Lake Champlain, which lies nearly north and south, dividing the States of Vermont and New York. It is about 150 miles in length, and measures fourteen miles at the point where it attains its greatest breadth. The banks of the lake are in general low and marshy, and for about twenty miles at its southern extremity, it assumes the appearance of a river, hardly affording sufficient space to permit a vessel to turn. This lake is navigable throughout its whole extent for vessels drawing five feet of water, and several fine steam-boats ply on it while the navigation is open. The principal towns on its shores are St John's, Plattsburg, Ticonderoga, Whitehall, and Burlington, at which last place the steam-boats for its navigation are built. It is connected with the river Hudson by the Champlain Canal, but it discharges its surplus water into the St Lawrence by the river Richlieu, called also the Sorell, on which the towns of St Dennis, St Charles, and Sorell, are situated. The chief trade of Lake Champlain consists in exporting iron-ore and timber; the iron is sent to New York by the canal, and the timber to the St Lawrence by the river Richlieu. Its waters are exceedingly pure, and are subject, during the wet seasons of the year, to great augmentation. The captain of the steamer by which I travelled informed me that, in the spring of 1816, when the snow was leaving the ground, the surface of the lake rose to the height of nine feet above its summer water level. Its navigation, like that of the other lakes, is suspended for five months in the year by ice, and transport is carried on during that period by sledges, which run on its surface.

CHAPTER III.

RIVER NAVIGATION.

The sizes and courses of the North American Rivers influenced by the Alleghany and Rocky Mountains.—Rivers flowing into the Pacific Ocean.—Rivers flowing into the Gulf of St Lawrence.—River St Lawrence.—Lakes, Rapids, and Islands on the River.—Lachine Canal.—St Lawrence Canal.—The Ottowa.—Rideau Canal.—Towing vessels on the St Lawrence.—Tides.—Freshets. Pilots, &c.—Rivers rising on the east of the Alleghany Mountains, and flowing into the Atlantic Ocean, and north-east corner of the Gulf of Mexico.—The Connecticut.—Hudson.—Delaware.—Susquehanna.—Patapsco.—Potomac, &c.—Mississippi and its tributaries.—The Yazoo.—Ohio.—Red River.—Arkansas.—White River.—St Francis.—Missouri.—Illinois, &c.—State of the Navigation.—“ Snags,” “ Planters,” “ Sawyers,” and “ Rafts.”—Construction of Vessel for removing “ Snags,” &c.

THE rivers of North America are no less interesting features in the hydrography of that country than her inland sounds and lakes; and the great lines of navigable communication which so many of them afford, extending in all directions from the shores of the ocean to the very heart of the country, and forming great public highways for the easy and quick transport of the most bulky produce of the interior, as well as the sea-borne manufactures and luxuries of foreign lands, entitle them, in a commercial point of view, to an equal share of attention.

It is impossible to convey to the reader an adequate idea of those vast bodies of moving water, or to describe the feelings which the traveller experiences, when, for instance, after crossing the Alleghany Mountains, and completing a fatiguing land journey from the eastern coast of several hundred miles into the interior of the country, he first comes in sight of the Ohio River at Pittsburg. Here, in the very heart of the continent of North America, the appearance of a large shipping port, containing a fleet of thirty or forty steamers moored in the river, cannot fail to surprise him; and his astonishment is not a little increased if he chances to witness the arrival of one of those steamers, whose approach is announced long before it makes its appearance by the roaring of its steam, and the volumes of smoke and fire which are vomited from the funnels; but his wonder only attains its height when he is told that this same vessel has come direct from New Orleans, in the Gulf of Mexico, and that fifteen days and nights have been occupied in making this inland voyage, of no less than two thousand miles, among the meanderings of the Mississippi and Ohio.

The continent of North America may be said to be divided into four distinct portions by the ranges of the Alleghany and the Rocky Mountains, which run from north to south, in directions nearly parallel to each other, and regulate the lengths of the various rivers by which the country is drained, and, as it were, assign to each, the quantity of water which is due to it, and the direction it must follow in its progress to the ocean. I shall consider the rivers, therefore, under four distinct heads. *First,* those which rise on the west of the Rocky Mountains, and flow into the Pacific Ocean. *Second,* those which take their rise to the north of the mountain ranges, and discharge themselves into the Atlantic Ocean by the river and gulf of St Lawrence. *Third,* those which have their sources on the east of the Alleghany Mountains, and discharge themselves into the Atlantic and the north-eastern part of the Gulf of Mexico; and, *fourthly,* the rivers comprehended under the head of the Mississippi and its tributaries, which have their rise in the great valley stretching between the Alleghany and the Rocky Mountains.

Our information respecting the rivers comprising the first of these divisions, or those which discharge themselves into the Pacific Ocean, is very limited, owing to the unexplored state of the country lying to the westward of the Rocky Mountains, through which they flow. It is certain, however, that their courses are short, as the base of the Rocky Mountains, which are said to be abrupt and lofty, extends to within a few hundred miles of the shore, a circumstance which renders it not unlikely also that the declivity of their beds is considerable, and their currents in general too rapid to admit of easy navigation. Those which have been visited are Frazer's River, the Caledonia, the Colorado, and the Columbia. The Rivers Colorado and Columbia, are said to be navigable for a considerable distance.

The rivers which flow into the great western lakes, and those joining the St Lawrence in its course from Lake Ontario to the sea, form the second division.

Although the St Lawrence does not assume its name until it issues from Lake Ontario, it nevertheless takes its rise to the westward of Lake Superior. Between Lakes Superior and Huron, it is called St Mary's river. From Lake Huron it flows under the name of the St Clair into the lake of that name, from whence to Lake Erie it is called the Detroit River, and between Lakes Erie and Ontario the Niagara; but still it is essentially the same stream, in the same way as the Rhone, both above and below the Lake of Geneva, is considered the same river, but there retains the same name. When viewed in this light, the St Lawrence may be said to have a course of upwards of two thousand miles, and to receive the waters of about thirty rivers of considerable size. After leaving Lake Ontario, it assumes the name of the St Lawrence, and receives, in its progress to the ocean, by the river Richlieu or Sorell, the water of Lake Champlain, and is also augmented by the streams of the Ottowa, St Francis, St Maurice, Chaudière, and Charles rivers.

Receiving the whole surplus waters of the North American lakes, and the drainage of a great tract of country traversed by the numerous streams which join it in its course to the ocean, the St Lawrence, as regards the quantity of its discharge, presents abundant advantages for safe and easy navigation. The stream of the upper part of the river, however, is much distorted by numerous expansions and contractions of its banks, and also by declivities or falls in its bed, and clusters of small islands, which render its navigation exceedingly dangerous, and in some places wholly impracticable for all sorts of vessels excepting the Canadian *batteaux*, which are strong flat-bottomed boats, built expressly for its navigation. In several parts of its course the river expands into extensive lakes; and in its waters, which are thus distributed over a great surface, numerous shoals occur, among which the ship-channel is generally tortuous and narrow, and only navigable in daylight. In some places again, the St Lawrence forces its way between high banks which encroach on its bed, and leave a comparatively narrow gullet for its passage, and in others it flows over a steep and rugged bottom. These sudden contractions and declivities interrupt the peaceful flow of the stream, and produce *chutes*, as they are there called, or rapids, some of which are wholly impassable for vessels of large size, and others can only be navigated in certain states of the tide. The islands, which occur chiefly in the upper part of the river between Montreal and Lake Ontario, also distort the channel, and give rise to rapids which are no less detrimental in a commercial point of view.

Notwithstanding the numerous impediments to navigation, occasioned by the form of its bed, the river St Lawrence, between Montreal and Quebec, presents a scene of constant animation and bustle, until the approach of winter causes a suspension of its trade; on its stream the whole exports of Upper and Lower Canada are borne to the ocean, and by its current the valuable timber of that country is floated from its native forests to Quebec, where it is shipped for exportation. After passing the island of Orleans (on which the great timber ship Columbus was built), is the city of Quebec, the first place of importance that occurs in ascending the St Lawrence. The banks of the river at this place are high and precipitous. The fort of Quebec, built on Cape Diamond, is elevated 350 feet above the surface of the water, and commands a view of the river and surrounding country, which, for extent and grandeur, is perhaps unequalled in any part of the world. The river St Charles joins the St Lawrence close to the town, and the Chaudière flows into it a few miles farther up.

The first obstacle to navigation are the Richlieu rapids, about eighty miles above Quebec, where the banks approach each other, and leave a narrow channel of only about half a mile in breadth, which contracts the vast body of water discharged by the river, and produces a current of such strength that vessels, unless aided by steam, have great difficulty in stemming it. The rapids extend over several miles, and sometimes, it is said, run with a velocity of six miles per hour, and, notwithstanding this, the water, from its great depth, presents a smooth and unbroken surface.

From the Richlieu Rapids to Montreal, the banks are low, and the country, for some distance on each side, is flat and monotonous ; and were it not for many beautiful villages, with their churches and polished tin spires which meet the eye in close succession, and tend to diversify and enliven the scenery, the sail from Quebec to Montreal would not prove very inviting. About mid-way between those places, the bed of the river expands, and at last attains the breadth of nine miles, forming the large sheet of water called Lake St Peter, which is twenty-one miles in length. In this lake there is very little current, and but a small depth of water, the natural consequence of the river being extended over a great surface. A deep channel winds through the middle of the flat, affording an intricate passage for vessels, which, in their progress through it, are compelled to cast anchor after sunset. The course of this narrow channel is marked by buoys ; and lights are exhibited at its two extremities for guiding vessels out of it which happen in the course of their voyage to reach either termination immediately after night has set in, in which case they are enabled to proceed on their course without encountering the delay of anchoring all night in the lake.

The rivers St Maurice and Richlieu or Sorell, flow into Lake St Peter. At the mouth of the St Maurice stands the town of Trois Rivieres, which contains about 3000 inhabitants, and ranks as the third town in Lower Canada. The Richlieu enters the lake at its southern extremity, and at its mouth stands the town of William Henry or Sorell. The Richlieu, as formerly noticed, flows from Lake Champlain, from which a great deal of timber is annually floated by its current to the St Lawrence.

About a mile below Montreal, the navigation encounters a great impediment in the rapids of St Mary, caused by St Helen's Island, which lies in the middle of the river. Here the current, it is said, runs with a velocity of six miles an hour ; and about fifteen years ago, before the powerful and well-constructed steam vessels which now navigate the St Lawrence were built, a relay of oxen was kept at this place for assisting the steamers to ascend the rapids. It is unfortunate for Montreal, nautically or commercially speaking, that it is situate above instead of below these rapids, as it renders the port difficult of access to all classes of vessels.

Montreal, as before noticed, is 180 miles from Quebec, and 580 miles from the Gulf of St Lawrence. It is at the head of the ship navigation ; and although upwards of 880 miles distant from the Atlantic Ocean, vessels of 600 tons ascend the river, and lie afloat at the quays.

The Lachine Rapids, extending over about seven miles of the river's course, lie immediately above Montreal. As the velocity with which the water runs at this place renders navigation impracticable, a work, called the Lachine Canal, has been executed, at an expense of L.115,000, in order to avoid this obstruction to the navigation. This canal was completed in the year 1824, and was the first work of the kind formed in Canada. It extends from Montreal to a place called Lachine, a distance of nine miles, and measures forty-eight feet in breadth at the water-line, twenty-eight feet at the bottom, and five feet in depth. The rise is forty-eight feet, which is overcome by means of six locks of eight feet lift each. The locks and other works on the line of the canal, which are subject to much tear and wear, and require strength and durability, are constructed of red sandstone, in a well-finished and substantial manner.

The St Lawrence is navigable from Lachine to the "Cascades," where the "Cedar Rapids" again stop its navigation for about sixteen miles. Above this the river expands, and forms the navigable lake of St Francis, which is twenty-five miles in length, and, in some places, attains the breadth of five and a half miles. The town of Regis stands at its northern extremity, and is inhabited by part of a large tribe of Indians, who have a settlement here on a tract of land granted to them by the British Government. Above Lake St Francis are the Longue Saut Rapids. These are nine miles in length, and flow with greater velocity than any of the others which have been mentioned. At their head stands the town of Cornwall, from which the river is navigable to Kingston, at the entrance to Lake Ontario. The towns of Ogdensburg, Prescott, and Brockville, are situate on the banks of the river, between Cornwall and Lake Ontario.

A few miles below Kingston is the celebrated "Lake of the Thousand Isles." At this part of its course, the St Lawrence assumes a great breadth, and its surface is thickly studded with islands, varying from a few square feet to several acres in extent. There are said to be upwards of 1500 of them in the lake; and, though they form an interesting and splendid object in the scenery of the river, they prove very detrimental to its navigation. A channel, having a sufficient depth of water for ships of the largest size, winds among the islands, and is in some places so narrow, that, when the wind is high, vessels have often difficulty in passing each other.

These obstructions in the St Lawrence are injurious to its character as a navigable river; but they impart to the scenery on its course a degree of grandeur and variety which is peculiarly pleasing to the traveller. In passing over some of the rapids which have been mentioned, the water is violently agitated and tossed into the air, covering the whole surface with a sheet of white foam, and forming a fine contrast to the clear blue of the untroubled part of the river. The fearless Canadians, however, daily descend these impetuous streams with their *batteaux* and rafts of timber, without encountering the least accident or inconvenience. The batteaux are strong flat-bottomed boats, well suited to the navigation of the rapids, and are generally manned by skilful navigators. They descend from Ogdensburg to Montreal, a distance of ninety-five miles, heavily laden with the produce of the country, and generally occupy about three days in making the voyage. Steam-boats ply regularly on those parts of the river which lie between the rapids ; but the batteaux, as formerly observed, are the only description of vessels that can, with any degree of safety, be taken over the rapids.

The province of Upper Canada has commenced a gigantic work, to supply these deficiencies in the navigation of the river, which is to be called the St Lawrence Canal. The first compartment of this work, extending from Cornwall, on the left bank of the St Lawrence, to a place called Dickinson's Landing, is twelve miles in length, and is intended to overcome the Longue Saut Rapids. This work was in a very advanced state when I visited the country. Two additional short canals, however, and an alteration in the dimensions of the Lachine Canal, must still be carried into effect, in order to complete the whole of the contemplated improvement, by which another communication, in addition to that already afforded by the Rideau Canal, will be opened between the lower part of the St Lawrence and the lakes. It is intended that the St Lawrence Canal shall have a breadth of 100 feet throughout its whole extent, and be capable of admitting the passage of all vessels under 100 feet in length, which do not draw more than eight feet of water. The locks are to be built of limestone, which is obtained in fine blocks and great abundance in the surrounding country.

The Ottowa, after a course of about 500 miles, joins the St Lawrence immediately above the island of Montreal. It is navigable to Bytown, 120 miles from its mouth; and the Grenville Canal, the locks and works connected with which have been formed on the same scale as those of the Lachine Canal, was constructed to obviate some of the rapids which occur on the river.

The Rideau Canal, leading from Bytown on the Ottowa to Kingston on Lake Ontario, was constructed by the British Government, chiefly with the view of providing a sheltered passage, at a secure distance from the frontier, for the transport of military stores to the lakes, in the event of war with the United States; and, notwithstanding its construction, a great deal of trade is still carried on by the batteaux which continue to navigate the rapids of the St Lawrence.

About seventy miles of the Rideau Canal consist of what is technically called *slackwater* navigation, which in this case is formed by damming up the waters of the Rideau river and lake, and increasing their depth so as to fit them for steamers of a pretty large size. The entrance of the canal at Bytown is 283 feet below Rideau Lake, which is the summit level, and 129 feet below Lake Ontario. There are several bold and arduous works on the line of this canal, the execution of which in so rough and unfavourable a country confers great credit on Colonel By, the principal, and Mr M'Taggart, the assistant, engineers, under whose directions they were conducted. The length of the canal is 135 miles; seventy miles of this, as before noticed, are slackwater navigation, and its cost is said to have been about L. 600,000. The works are constructed on a scale sufficient to admit vessels of 125 tons burden. It is much to be regretted that the locks of the Lachine Canal at Montreal had not been originally constructed of wood instead of stone, as in that case they might have been enlarged at a small cost, and rendered suitable for the same class of vessels which now navigate the Rideau Canal, the locks of which are of much larger dimensions, and consequently admit larger craft.

The Lachine Canal, the Rideau Canal, and the Welland Canal, constructed by the British subjects, together with Ohio Canal, constructed by the inhabitants of the United States, amount in all to four hundred and fifty-one miles in extent. These interesting works connect the Gulfs of St Lawrence and Mexico by a water communication, forming with Lakes Ontario and Erie, and the rivers St Lawrence, Ohio, and Mississippi, a gigantic line of inland navigation upwards of three thousand miles in length.

Vessels bound for Montreal are generally towed up the river from Quebec by large and powerful steam-boats, belonging to the "St Lawrence Steam-boat Tow Company." The company's charge for towing a vessel of 20 feet beam and 9 feet draught of water, from Quebec to Montreal, is L. 33 : 6 : 8, and for a vessel of 28 feet beam and 15 feet draught of water (the largest size that ever penetrates so high as Montreal), the charge is L. 83, 4s. Vessels of intermediate sizes are charged proportionally.

The art of towing vessels by steam-tugs is practised very extensively, and has been brought to great perfection both on the Mississippi, as formerly noticed, and on the St Lawrence. In both of these rivers the narrowness of the navigable channels, and the great distance at which the ports are removed from the sea, render some other means than sails, for propelling the vessels navigating them, absolutely necessary. The most powerful tow-boat on the St Lawrence when I visited the country was the "John Bull." By this vessel I passed from Quebec to Montreal, a distance of 180 miles, in forty hours, being at the rate of four and a half miles an hour, against a current averaging about three miles an hour. Upon this occasion she had no fewer than five vessels in tow ; one of these drew twelve and a half, another ten and a half, two of them drew nine, and the fifth about seven feet of water. The vessels were all towed by separate warps, and were ranged astern of each other in two lines, three of them being made fast to the larboard, and two to the starboard side of our vessel. The management of a steamer with so great a fleet of vessels in tow, in the intricate navigation and strong current of the St Lawrence, requires no small degree of caution and skill on the part of the captain, who on this occasion had his whole charge most perfectly under command: when it was necessary to stop the steamer's progress for the purpose of taking in fuel or goods, he dropped the vessels astern, and picked them up again on resuming his course with the greatest dexterity. Captain Vaughan, who commands the "John Bull," informed me, that it is by no means uncommon, at certain seasons of the year, to have six vessels in tow, and from 1200 to 1500 passengers on board of his vessel at the same time. He tows every vessel by a separate line, and generally keeps them all astern in preference to taking any of them alongside of the steamer, an arrangement which, in the St Lawrence, where the navigable channel is in many places very contracted, and often impeded by large rafts of timber, would be very apt to occasion accidents.

There is a rise of twenty feet at spring-tides at the quays of Quebec; and when there is not much flood-water in the river, it is said to be affected by high tides to the distance of fifty miles above the city, or about 750 miles from the Atlantic Ocean at the entrance of the Gulf of St Lawrence. The floods or freshets, which occur at the breaking up of winter, are chiefly caused by the melting snow, and occasion a periodical rise in the surface of the river, which is sometimes from this cause raised as much as ten feet above its summer water level. When I visited the St Lawrence in May 1837, it was under the influence of a freshet produced by the melting of the snow; and it was said to have raised the river to a greater height than had ever been known before, the water being at that time several feet above the level of some of the quays in Montreal. Mr M'Taggart, who had a good opportunity, during his residence in Canada, of making accurate observations, states, that the whole quantity of water annually discharged into the sea by the St Lawrence may be estimated at 4,277,880 millions of tons, and also, that the quantity of water annually discharged *into* the St Lawrence from the melting of the snow may amount to 2,112,120 millions of tons. As the whole of this great body of water is poured into the stream in a short space of time, it materially affects the level of the water, causing it to overflow the banks, and cover every low lying tract of ground in the vicinity of the river.

The severe and protracted winter of Canada, so hostile to the interests and prosperity of the country, puts a stop to the navigation and trade of the St Lawrence for at least four and a half months annually, and during great part of that period the ice at Quebec often forms a spacious and safe bridge across the river.

The navigation of the Gulf of St Lawrence, through which the river discharges itself into the Atlantic, is very hazardous. In addition to the dangers arising from the masses of ice which are constantly to be met with, floating on its surface, for nearly one-half of the year, it is subject to dense and impenetrable fogs, and its rocky shores and desolate islands afford neither comfort nor shelter to the shipwrecked mariner. One of the most desolate and dangerous of the islands in the Gulf, is Anticosti, which lies exactly opposite the mouth of the St Lawrence, and is surrounded by reefs of rocks and shoal water. Two lighthouses have been erected on it, and also four houses of shelter, containing large stores of provisions, for the use of those who have the misfortune to be shipwrecked on its inhospitable shores.

The lighthouses, buoys, and pilots, belonging to the St Lawrence, are under the control of the Trinity House of Quebec. The lights are by no means so numerous or efficient as the dangerous and crowded navigation of the river requires. There are ten lighthouses between Montreal and Anticosti, a distance of 580 miles, and these are nightly illuminated while the navigation of the river is open. The number of the licensed pilots is about 250, who are compelled to serve an apprenticeship, and to make at least one trip across the Atlantic previously to obtaining a licence to act in this capacity.

The rivers belonging to the third division, which take their rise on the east of the Alleghany Mountains, and flow into the Atlantic Ocean and the north-east corner of the Gulf of Mexico, are upwards of one hundred in number. They are distributed over the whole eastern part of the country; and, notwithstanding the shortness of their courses, extending only from the sea-coast to the base of the Alleghany Mountains, they afford an aggregate amount of upwards of 3000 miles of ship and boat navigation. The following are the most important of these streams.

The St Croix is a short river, having a course of about sixty miles, and is remarkable only as being the boundary between the United States and the British dominions in North America.

The Penobscot has a course of about 300 miles, and flows into the sea at Penobscot Bay in the State of Maine. It is navigable, for vessels of large burden, to the town of Bangor, which is situate fifty miles from the sea at the head of tide-water. Large quantities of valuable timber are annually exported from the towns on this river and bay.

Kennebeck River is the outlet of a small sheet of water called Moosehead Lake; it flows into the sea at Augusta in the State of Maine, after a course of about 230 miles, and is navigable for a distance of forty miles from the sea.

The Merrimac, flowing into the sea at Newburgh Port in Massachusetts, has a course of upwards of 200 miles, but, in consequence of several falls which occur in its bed, is navigable only for a distance of twenty miles from the sea. It affords very valuable water-power, and on its banks is situate the large manufacturing town of Lowell.

The Thames falls into Long Island Sound at New London and is navigable to the town of Norwich, fifteen miles from its mouth.

The Connecticut, after a course of 450 miles through a highly cultivated and fertile country, discharges itself into Long Island Sound. It is navigable for steamers and vessels of large burden to Hartford, a distance of forty miles, and by means of some short canal works, for steamers of a small size to Barnet in Vermont, which is upwards of 250 miles from the sea.

The Hudson rises in the neighbourhood of Lake Champlain, and pursuing an almost straight course of about 250 miles in a southerly direction, flows into the sea at the city of New York. Although that portion of the Hudson which is strictly a river, or in which the tide does not act, is by no means so remarkable for its size as many others in the United States, yet it is very interesting to the traveller, as well on account of the beauty of its scenery, as the importance and extent of its trade; and in this respect it holds a very high rank among the American rivers. It passes through a beautiful and sheltered tract of country, and the populous towns of Newburgh, Hudson, Albany, and Troy, and the military college of West Point, stand on its banks. The produce of the large State of New York and the great western lakes, as well as the imports for the supply of an extensive and populous district of the United States, are borne to and from the harbour of New York by the Hudson, and a large fleet of vessels is constantly engaged in its navigation.

This river is navigable, for ships of large burden, to the town of Hudson, about 120 miles from New York, and for vessels of smaller draught of water to Troy, about forty-four miles farther. By means of the Erie, Oswego, and Champlain canals, it is connected with Lakes Erie, Ontario, and Champlain. A large part of the trade of the river Hudson is carried on by sailing vessels of about 150 tons burden, having a great breadth of beam, and carrying masts of from 90 to 100 feet in height. These vessels, being dependent on the state of the winds, make tedious and uncertain voyages; but many of them, notwithstanding the introduction of steam-navigation, still enliven the river scenery with their white sails. The transport of goods, however, is now more generally carried on in large barges, towed by steamers which are exclusively devoted to this trade, as passengers go only by the larger and swifter boats built expressly for the purpose. The current of the Hudson is said to average about two and a half miles an hour, and the influence of the tide extends as far as Albany, 150 miles above New York. The only obstacle to navigation occurs a little below Albany, where there is a considerable shoal, called the Overslaugh, caused by several small islands lying in the fairway of the river. It is, however, at present passable for vessels drawing five or six feet of water, and is still capable of being much improved.

The Delaware has a course of about 310 miles, and falls into Delaware Bay near Newcastle ; it is navigable, for vessels of the largest class, for forty miles, to Philadelphia. From Philadelphia it is navigated by sloops, for a distance of thirty-five miles, to Trenton, which is at the head of tide-water, and above this it is navigable for boats of nine tons, which ascend the river about one hundred miles farther into the interior.

The Susquehanna flows into Chesapeake Bay. It is the largest river in the productive State of Pennsylvania, but is more celebrated for the beauties of its scenery than the facilities it affords for communication. Excepting for about five miles from its mouth, the navigation is completely stopped by the rugged and shelving formation of the rocky bed in which it flows. The course of this river is about 460 miles, and works are now in progress, for the improvement of its navigation, by the formation of short canals, and the construction of dams, so as to form an extensive line of slackwater navigation.

The Patapsco discharges itself into Chesapeake Bay, and is navigable, for vessels drawing eighteen feet of water, to Baltimore, which is at the head of tide water, and is about fourteen miles from Chesapeake Bay. The whole course of the Patapsco is only about one hundred miles.

The Patuxent rises to the west of Baltimore, and

flows into Chesapeake Bay. It has a course of about one hundred miles in length, and is navigable to the distance of sixty miles from its mouth.

The Potomac has its source in the Alleghany Mountains, and is 335 miles in length. It is seven and a half miles in breadth at its entrance into Chesapeake Bay, and is navigated, by vessels of the largest class, as far as Washington, the seat of government of the United States, which is situate about 103 miles from its mouth. The tide flows three miles above Washington, but beyond this point the river is obstructed by shoals, and several short canals have been constructed for the improvement of its navigation.

The Rappahannock has a course of 176 miles, and is navigable to the town of Fredericksburg, about 110 miles from its junction with Chesapeake Bay.

York River also flows into Chesapeake Bay, and has a course of one hundred miles, thirty miles of which are navigable for large vessels.

The James River has a course of upwards of 400 miles, and discharges itself into the Atlantic, at the southern extremity of Chesapeake Bay. It is navigable, for vessels of 125 tons burden, to the town of Richmond, situate 122 miles from its mouth, where the navigation is obstructed by falls in the river. By means of a canal which has been formed to overcome this obstacle, batteaux are now enabled to ascend the river to a distance of 352 miles from the sea.

The Roanoke flows into Albemarle Sound in North

Carolina, after a course of 370 miles. It is navigable, for vessels of forty-five tons, to Halifax, seventy miles. Batteaux ascend the river to the distance of 300 miles from its mouth.

The Pamlico falls into Pamlico Sound. It has a course of 200 miles, and is navigable for forty miles.

The river Neuse has a course of 271 miles; Cape Fear, 288; Pee Dee, 415; Santee, 370; and Edisto, 161 miles. These rivers are in North and South Carolina, and are said to be capable of affording, by means of some small improvements, about 630 miles of boat-navigation.

The rivers Ashley and Cooper in South Carolina, have courses of forty-three and forty-four miles, and, at their junction, form the harbour of Charleston.

The Savannah River flows between the states of South Carolina and Georgia. It has a course of 340 miles, and is navigable, for vessels of the largest size to the town of Savannah, situate eighteen miles from the sea. Above this, steam-navigation extends as far as Augusta, 140 miles.

The great Ogeetchee is navigated by small vessels for 300 miles, the Alatamaha for 220, the Santilla for 180, and the St Mary for 150 miles from the sea. The rivers St John and Suwanee, in Florida, are said to have courses of about 250 miles. Many of the streams in the southern part of the United States, however, and more particularly in Florida, have never been fully explored.

The Appalachicola has a course of 425 miles. It is formed by the junction of the Chattahoochee and Flint rivers, and discharges its waters into the Gulf of Mexico. It is navigated by steamers to the town of Columbus, 160 miles from its mouth.

The Mobile river is formed by the junction of the Alabama and Tombeckbee. The Alabama has a course of 500, and the Tombeckbee of 350 miles. The Alabama affords ship-navigation to Clairbone, 100 miles, and batteaux-navigation to Fort-Jackson, 200 miles. The Tombeckbee is navigated by ships as far as St Stephens, 100 miles, and by boats to the falls of the Black Warrior, 250 miles from the Gulf of Mexico.

The part of North America which extends from north to south between the great northern Lakes and the Gulf of Mexico, and from east to west between the ranges of Alleghany and Rocky Mountains, includes within its limits the valleys of the Mississippi, Missouri, and Ohio, and is remarkable for the extreme richness and fertility of its soil, which, after being brought into cultivation, yields with little labour a very abundant harvest. These fertile valleys include nine of the United States of America, and a great part of them is now in a high state of cultivation, and thickly peopled. In the State of Louisiana, the crops grown are sugar, cotton and tobacco ; and in Mississippi and Arkansas cotton is produced in great abundance, and of fine quality. Tennessee affords cotton and tobacco, and Kentucky produces hemp, tobacco, wheat and Indian corn. The states of Ohio, Indiana, Illinois, and Missouri, are too far removed from the equator for the growth of cotton, sugar, or tobacco, and their inhabitants confine all their attention to raising grain. The geographical structure of North America shuts up this immense tract of land from any direct communication with the seas which wash its eastern and western coasts; for if we trace upwards, in their course of many hundred miles through the eastern States, those numerous large navigable rivers which discharge themselves into the Atlantic, we find them holding the character of rivulets long before we penetrate even to the verge of these fertile valleys; and on the western coast of the country, the range of the Rocky Mountains extending along the shores of the Pacific, present an insurmountable barrier to any direct communication with that ocean.

The Mississippi, however, and its numerous navigable tributaries, afford a perfect and easy access to the remotest corners of these States. The produce which annually descends the river, is valued at the enormous sum of fourteen millions of pounds Sterling, and its four mouths pour into the Gulf of Mexico the drainage-water of a district of country which has been estimated at no less than 1,300,000 square miles in extent. The source of the Mississippi is said to have been discovered in the year 1832. It is situate to the westward of the great lakes, at a distance of upwards of three thousand miles from the Gulf of

Mexico, and at an elevation of fifteen hundred feet above its surface. The river flows from its source as a small stream, and gradually gathering strength, precipitates itself over the falls of St Anthony, after which it swells in importance at every step of its course, gaining accessions of strength from the numerous small rivers which pour in their tributary streams from all directions, until it is at length joined by the great Missouri. The character of its waters, formerly clear and tranquil, is here completely changed, and the combined streams of the two rivers flow on in a deep and muddy torrent. The Ohio, the Arkansas, the Red River and many other large streams, fall into this giant of rivers, which, swelled by the waters of its various tributaries, whose aggregate length is upwards of forty-four thousand miles, at last pours itself into the Gulf of Mexico.

The Mississippi, exclusively of its tributaries, affords an uninterrupted line of navigation for 2250 miles between its mouth and the falls of St Anthony. New Orleans, the most important town on the river, has already been noticed. The town of Natchez, which is about 380 miles from its mouth, stands on the left bank ; it is a place of considerable importance, and is the highest point visited by sailing vessels ; above this, the Mississippi is navigated only by steam-boats. St Louis, on the right bank of the river, about eighteen miles below its junction with the Missouri, is also a place of great trade.

The Mississippi forms a striking contrast to the St Lawrence, which, as has been already observed, flows in a rocky bed, occasionally expanding into extensive flats, or contracting its limits, and thus presenting great impediments to navigation. The bed in which the Mississippi flows, is of a soft alluvial formation, maintaining a nearly uniform breadth throughout its whole course, and affording, at every point below the falls of St Anthony, a sufficient depth of water for vessels of the largest size. Its breadth from the Gulf of Mexico to its junction with the Missouri, which is a distance of about 1100 miles, is said to be no more than half a mile, and its average depth no less than a hundred feet. The principal mouth of the Mississippi has a bar, on which the depth of water in 1722, according to Malte Brun, was twenty-five, in 1767 twenty, and in 1826 sixteen feet. Captain Hall mentions that in 1828 it was only fifteen feet. The vast tract of Delta land at the mouth of this river, caused by the deposition of the earthy matter carried down by its current, is gradually extending its limits, and stretching into the Gulf of Mexico, a circumstance which has led some to remark that, in the course of time, the whole Gulf of Mexico, at present occupied by the sea, may be filled up by these alluvial deposits, and become a flat plain watered by an extension of the Mississippi.

In enumerating the tributaries of the Mississippi, I shall first notice those flowing into it from the east, and afterwards those which have their rise in the western country, in the order in which they occur in ascending the river.

The Yazoo, flowing through the State of Mississippi, joins the river about 450 miles from the sea, and is navigable for 150 miles.

The Ohio, the largest of its eastern tributaries, is, excepting at one or two parts of its course, a smooth running stream. It is formed by the junction of the Monongahela and Alleghany rivers. The Monongahela is navigable, for small boats, for two hundred miles, and the Alleghany is navigable, for boats of ten tons, for two hundred and sixty miles. The Ohio flows over a course of 945 miles, and discharges itself into the Mississippi about 1000 miles above New Orleans. Its banks, which rise rather precipitously, are thickly covered with fine timber, and the country through which it passes is highly cultivated, and very productive. The navigation of the river is stopped for about four months every year by ice. The principal towns on the Ohio are Louisville, Cincinnati, Wheeling and the manufacturing town of Pittsburg, which stands at the head of the navigation, on a point of land formed by the junction of the rivers Monongahela and Alleghany.

During the spring months, when the Ohio is swollen, steam-boats of the largest class, drawing from eight to ten feet of water, ascend from the Gulf of Mexico to Pittsburg, a distance of nearly 2000 miles. But when the water is low, steamers cannot ascend higher than Louisville, in Kentucky, which is situate on the left bank of the river, 560 miles below Pittsburg. Here the river has a fall, occasioned by an irregular ledge of limestone rock, of twenty-two feet six inches in two miles, which produces rapids that can only be passed when the river is high. The Louisville and Portland Canal, constructed with a view to remove the obstruction to navigation occasioned by this fall, is rather more than two miles in length, and is excavated in rock nearly throughout its whole extent. It is sixty-eight feet in breadth, and sixteen feet in depth, affords a passage for all steam-boats under 180 feet in length, and is used by them when the low state of the water in the river renders the rapids impassable.

The canal has three lift-locks, measuring 183 feet in length, and 50 feet in breadth, and one guard-lock measuring 190 feet in length, and 50 feet in breadth, all of which are built of stone.

Several shoals occur, in the upper part of the river, which are also very hurtful to the navigation, as the current on many of them runs with considerable velocity. In ascending the Ohio, the steamer by which I travelled was very deeply loaded, and we were detained several hours in attempting to pass one of these shoals called the "White Ripple." Many unsuccessful efforts were made, but the power of the engines could not surmount the obstacle, until some of the crew ascended the stream in a boat, and dropped an anchor with a strong cable attached to it, in the middle of the channel ; the other end of the cable was made fast to the capstan of the steam-boat, and the vessel was at length, after much labour and detention, warped through the rapid.

The principal tributaries flowing into the Ohio from the north, are the Muskingum, which is navigable for 120 miles, the Miami, navigable for 75 miles, the Scioto, which is navigable for 120 miles, and the Wabash. The Tennessee river flows into the Ohio from the south. It is 850 miles in length, and is navigable to Florence, a distance of 250 miles. At this place there is an expansion in the bed of the river ; and a collection of stones, called "the Mussel Shoal," terminates the navigation. The other tributaries flowing into the Ohio from the south, are, the Cumberland river, navigable for 440 miles, Green river, for 150 miles, Kentucky river, for 130 miles, and Licking river, for 70 miles. The aggregate length of the Ohio and its tributaries is about seven thousand three hundred miles.

The Illinois enters the Mississippi about 160 miles above the Ohio, and is navigable for steam-boats for about two hundred miles.

Ouisconsin and Chippewa rivers take their rise in the neighbourhood of the lakes, and are both navigable for some distance.

The most southern tributary of any importance which flows into the Mississippi from the west is the Red River. This river takes its rise at the base of the Rocky Mountains, and is 1500 miles in length; but its navigation is obstructed by a huge pile of wood, composed of large trees, which having been swept away in floods, and floated down the stream, have finally found a resting place in the bed of the river, among their former neighbours of the forest. This obstruction, which is called the "Red River Raft," has been accumulating for ages. It commences about 500 miles above its mouth, and is said to extend about seventy miles. Measures have been adopted for effecting its removal, and should this arduous undertaking, which is at present in progress of execution, be successful, the navigation of the river will be extended 500 miles farther into the interior of the country. The Washita, one of the tributaries of the Red River, has a course of 450 miles.

The Arkansas has its source in the Rocky Mountains, and is said to be upwards of 2500 miles in length, and with its tributaries 4500 miles. Steamers can ascend this river for 640 miles from the Mississippi.

The White River, after a course of upwards of 1200 miles, including its tributaries, flows into the Mississippi, twenty miles above the Arkansas, and is navigable for 400 miles.

The St Francis has a course of 450 miles, but its entrance is choked by a large stationary raft of drift timber which puts an effectual stop to the navigation of the river.

The Merrimeg is navigable for 200 miles.

The Missouri joins the Mississippi 18 miles above the town of St Louis, and about 1200 miles from the Gulf of Mexico. It is, in every respect, the greater of the two rivers, but the Mississippi having been first discovered, the original name has been retained. The sources of the Missouri are in the Rocky Mountains, its whole course is 3217, and in connection with all its tributaries upwards of 10,000 miles. Its navigation is uninterrupted for 2532 miles from its mouth, and is there broken by the falls of the Missouri, which are said to vie in grandeur with those of Niagara, but the river is navigable above the falls for 500 miles.

The lead mines on the river Missouri, are of very great value. The district in which the lead occurs is about seventy miles in length, and forty-five miles in breadth The government of the United States have reserved 150,000 acres of land in the state of Missouri as government property. This is let in small lots to persons who undertake to open the mines ; they are now very extensively worked, and a large quantity of lead is prepared on the spot, and brought down the Missouri for the market.

The tributaries of the Missouri are the Gasconade, navigable for 150 miles; the Osage, said to be navigable for 500 miles ; the Chariton for 300 miles, the Tauzas for 200, and the Yellowstone for 800 miles.

The Moine flows into the Mississippi, 130 miles

above the Missouri, and is supposed, with its tributaries, to be navigable for a distance of 1500 miles.

The St Peter's, which is the most northern of the tributaries, has a course of 500 miles, and is navigable only for boats.

With the exception of the falls at Louisville, and the White Ripple on the upper part of the Ohio River, the Mississippi, and the navigable tributaries which have been enumerated, are perfectly free from those obstructions to navigation which are caused by any irregular formation in the beds or banks of the stream. Their currents have been estimated to run at the average rate of about three miles an hour. In some places, shoals or rapids occur, but these are by no means formidable, and do not affect the passage of steamers to a greater extent than by retarding their progress a little in ascending the river. Some dangers, however, exist, which are peculiar to the navigation of the western waters of America, and are even more to be dreaded than currents and rapids produced by permanent obstructions in the bed of the stream, as they are constantly changing their positions, and springing up afresh every day, so that they cannot be guarded against by any previous knowledge of the navigation of the river. These dangers are caused by large trees, which, being precipitated into the water, by the river undermining its banks, are borne away on the current, and occasionally get entangled, and even become firmly fixed, in the bed of the stream. Sometimes a branch of the tree is seen projecting from the water, but often no part of it is visible, and then the only indication of the existence of these hidden dangers is a slight ripple on the surface of the water. They have received from the boatmen of the Mississippi, the names of "Snags," "Planters" and "Sawyers," bearing one or other of these designations, according to their positions and the manner in which they are fixed in the river. The term "snag" is applied to a tree firmly imbedded in the bottom, and lying at a considerable angle, with its top inclined down the stream. A "planter" is a tree firmly fixed in a perpendicular position; and a "sawyer" is the name applied to a tree whose roots or branches have become entangled in the bed of the river, and, whose trunk being loose, is kept constantly swinging up and down by the current, alternately shewing its head, and plunging it under the surface. Sometimes several of these trees collect together in the same place, and form a small islet, which, after maintaining its position for some time, and gradually increasing its dimensions, at length attains an enormous magnitude, and often becomes an impassable barrier, extending along the river's course for many miles. This is what the boatmen call a "raft." It generally occurs in the tributaries of the Mississippi, and not in the river itself. One instance of this is afforded by the Red River, already mentioned, and another by the Atchafalaya, a river flowing *out* of the Mississippi, at a point about 250 miles from the sea. The Atchafalaya raft, which is particularly noticed in Captain Hall's work on North America, extends over a space of twenty miles ; but the river's bed, for the whole of this distance, is not filled up with drift timber ; the actual length of the raft itself being only about ten miles. The Atchafalaya is 220 yards in width, and the raft extends from bank to bank, and is supposed to be about eight feet in thickness.

All these obstructions are most injurious to the navigation of the Mississippi and its tributaries, and have, on many occasions, caused great loss both of lives and property by sinking steamers. The "snags" are more dangerous than any of the other obstructions. They are generally encountered by vessels on their upward passage. Vessels descending the river keep in the middle of the stream, where the water is deep and the current is strongest, while those ascending the river keep as close to the shore as possible, where they have a more gentle current and shoaler water, and, of course, are more apt to be injured by impediments in the bottom. Besides, as the "snags" are always inclined down the stream, vessels, going in the direction of the current, slide easily over them, if they happen to come in contact with them ; but their inclined position renders them exceedingly dangerous for vessels ascending the river, which obviously encounter them in their most destructive position. The strongest vessels in the western waters are unable to withstand the shock occasioned by running against a "snag." It almost invariably pierces their bows, when they generally fill with water and go down. Sevėral steamers are built with false bows, called " snag-chambers," as a palliative of the danger arising from accidents of this kind. In the event of the bow being stove in, the small compartment called the " snag-chamber," in the fore part of the vessel, is all that is filled with water, and her buoyancy is thus very little affected.

Some grants of money have been voted by the government of the United States for the improvement of the western water navigation. The money has been expended in removing, from different parts of the Mississippi and its tributaries, the stationary rafts of timber and snags by which their streams are obstructed. For this purpose, an apparatus called a " snag-boat," has been used with much success. The machine consists of two hulls, firmly secured to each other, at a distance of a few feet apart ; and over the intervening space a deck is thrown, having an aperture left in the centre. A powerful crab is placed over this aperture, from which strong chains and grapplings are suspended in the space between the two vessels. The " snag-boat" is propelled by paddle-wheels, which, with the gearing for raising the snags, are worked by a steam-engine placed on its deck. In using the apparatus, the vessel is brought to an anchor over the snag or obstacle to be removed, and the grapplings are made fast to the pieces which are to be raised. The paddle-wheels being thrown out of gear, the engine is applied to work the crab, by which the snag is torn from its hold in the bottom of the river, and, after being cut in short pieces, is allowed to float down the stream. This "snag-boat" has been extensively used on the Red River, in the partial removal of the large stationary raft formerly noticed, which at present obstructs the navigation of the stream.

The Mississippi and Ohio rivers are perfectly pure and limpid ; but after being mingled with the water of the Missouri, which holds a large quantity of alluvial matter in suspension, they assume a red and muddy appearance. A quantity of water, taken from the lower part of the Mississippi, and allowed to settle for fifteen or twenty minutes, deposits a thick cake of mud on the bottom of the vessel containing it ; but, notwithstanding this, the water is supposed by many persons to be healthful, and, after undergoing the process of filtration, is very generally used for domestic purposes by the inhabitants of all the towns situate on the river.

The average height of the annual rise in the waters of the Ohio is fifty feet, the lowest state of the river occurring in September, and the highest in March ; but I was informed that the waters of the Mississippi and Missouri are not subject to so remarkable a change of level.

The following interesting details are from Captain Hall's work on North America, which contains much valuable information regarding the Mississippi.

"At New Orleans, the difference between the level of the highest water and that of the lowest is thirteen feet eight inches perpendicular, English measure. The sea is distant from the city upwards of 100 miles, and, as the tide is not felt so far, the rise and fall alluded to are caused exclusively by the rainy and dry seasons in the interior."

"In proportion as we ascend the river, we find the perpendicular space between the rise and fall of its surface to increase. Near the efflux of the river Lafourche, the rise and fall is twenty-three feet. This is about 150 miles from the sea. At a place called Baton Rouge, 200 miles from the sea, the pilot-books state the perpendicular rise and fall of the river at thirty feet. At Natchez, which is 380 miles from the sea, it is said to be fifty feet. After it has flowed past Natchez, the volume of water in the Mississippi is dissipated over the Delta by such innumerable mouths, and overflows its banks at so many places, that the perpendicular rise and fall is of course much diminished. The velocity of the middle current seldom exceeds four miles an hour any where between the confluence of the Ohio and the sea.

"The width of the river at New Orleans at low water is 746 yards, which is somewhat less than half an English statute mile, being very nearly four-tenths, —the mile being 1760 yards. At high water it is 852½ yards broad, or 106½ more than at low water.

This, however, is still under half a mile, being a little more than forty-eight hundredths.

"I am the more particular in stating these measurements, from high authority, because a general belief prevails, I think, that the Mississippi is much broader. It may be mentioned that this river is fully as wide,—I should say rather wider,—abreast of New Orleans, than it is any where else from its mouth to the confluence of the Missouri, a distance of more than 1200 miles. During the whole of that extent, it preserves the most wonderful uniformity in width, very seldom, indeed, varying more than a hundred yards or so, over or under four-tenths of a mile. Mr Darby, in his very interesting description of Louisiana, at page 125, says:—'From careful triangular measurements of the Mississippi, made at Natchez,—at the efflux of Atchafalaya,—the efflux of the Plaquemine, near the efflux of the Lafourche,—at New Orleans,—Fort St Philips,—and at the Balize, the medial width was found to be short of 880 yards, or half a mile.' 'Eight hundred yards,' he adds, 'may be safely assumed as the width of the cubic column of water contained between the banks of the Mississippi.'*

"It is the depth which gives this mighty stream its sublimity. At New Orleans, the greatest depth observable at high water is 168 feet, but this is only at one place. At other parts, it varies much according

* A Geographical Description of Louisiana, by William Darby, Philadelphia, 1816.

to the deposits, and at some places is not fifty feet in depth. At Natchez, nearly 300 miles above New Orleans, when the water is at the lowest, I understand it is not less than seventy feet deep; and during that season the navigation of the river is exceedingly embarrassed by shoals, or bars, as they are called, which extend to a great distance off the points. Mr Darby, at page 135, gives the details of some measurements of the depth of the Mississippi, a little below the efflux of the river Lafourche, which I think is about fifty or sixty miles above New Orleans. He makes the depth there one hundred and thirty feet."

The level of the land on the banks of the Mississippi, for some distance before it discharges itself into the sea, is considerably below that of the surface of the river. Extensive embankments, similar to those of Holland and Belgium, have been erected for its protection, and form a continuous line on both sides of the river from New Orleans to St Francisville. Above this, and all the way to Natchez, which is about 380 miles from the sea, they occur only at intervals, where the flatness of the land has rendered their erection necessary. Captain Hall, on this subject, says: "The swollen river looked so like a bowl filled up to the brim, that it seemed as if the smallest shake, or the least addition, would send it over the edge, and thus submerge the city. The footpath on the top of the levée or embankment was just nine inches above the level of the stream. The colour of the water was a dirty, muddy, reddish sort of white, and the surface everywhere strongly marked with a series of curling eddies or swirls, indicative, I believe, of great depth."

These embankments, or *levées* as they are termed, are composed entirely of earth. They are from five to fifteen feet in height, and are made of sufficient breadth at the top to allow of a footpath being formed on them. They occasionally yield to the pressure of the river when in a flooded state, and give vent to its water, which on such occasions never fails to overflow and lay waste a large portion of the adjacent country.

CHAPTER IV.

STEAM NAVIGATION.

Introduction of Steam Navigation into the United States—Difference between the Steam Navigation of America and that of Europe —Three classes of Steamers employed in America — Eastern Water, Western Water, and Lake Steamers—Characteristics of these different classes—Steamers on the Hudson—Dimensions of the "Rochester"—Construction of the Hulls of the American Vessels—Arrangement of the Cabins—Engine Framing—Engines—Beams—Mode of Steering—Rudder—Sea-Boats—Dimensions of the "Naragansett"—Cabins—Engines—Paddle-Wheels—Boilers —Maximum speed of the "Rochester"—Power of the Engines—Mississippi Steamers—Their arrangement—Engines—Boilers—Lake Steamers—St Lawrence Steamers—Explosions of Steam-Boilers—Table of the Dimensions of several American Steamers.

WHATEVER differences of opinion may exist as to the actual invention of the steam-boat, there is no doubt that steam navigation was first fully and successfully introduced into real use in the United States of America, and that Fulton, a native of North America, launched a steam-vessel at New York in the year 1807; while the first successful experiment in Europe was made on the Clyde in the year 1812, before which period steam had been, during four years, generally used as a propelling power in the vessels navigating the Hudson.

The steam navigation of the United States is one of the most interesting subjects connected with the history of North America, and it is strange that hitherto we should have received so little information regarding it, especially as there is no class of works, in that comparatively new and still rising country, which bear stronger marks of long continued exertion, successfully directed to the perfection of its object, than are presented by many of the steam-boats which now navigate its rivers, bays, and lakes.

It would be improper to compare the present state of steam navigation in America with that of this country, for the nature of things has established a very important distinction between them. By far the greater number of the American steam-boats ply on the smooth surfaces of rivers, sheltered bays, or arms of the sea, exposed neither to waves nor to wind; whereas most of the steam-boats in this country go out to sea, where they encounter as bad weather and as heavy waves as ordinary sailing vessels. The consequence is, that in America a much more slender built, and a more delicate mould, give the requisite strength to their vessels, and thus a much greater speed, which essentially depends upon these two qualities, is generally obtained. In America the position of the machinery and of the cabins, which are raised above the deck of the vessels, admits of powerful engines, with an enormous length of stroke being employed to propel them; but this arrangement would be wholly inapplicable to the vessels navigating our coasts, at least to the extent to which it has been carried in America.

But perhaps the strongest proof that the American vessels are very differently circumstanced from those of Europe, and therefore admit of a construction more favourable for the attainment of great speed, is the fact that they are not generally, as in Europe, navigated by persons possessed of a knowledge of seamanship. In this country steam navigation produces hardy seamen, and British steamers being exposed to the open sea in all weathers, are furnished with masts and sails, and must be worked by persons who, in the event of any accident happening to the machinery, are capable of sailing the vessel, and who must therefore be experienced seamen. The case is very different in America, where, with the exception of the vessels navigating the Lakes, and one or two of those which ply on the eastern coast, there is not a steamer in the country which has either masts or sails, or is commanded by a professional seaman. These facts forcibly shew the different state of steam navigation in America, a state very favourable for the attainment of great speed, and a high degree of perfection in the locomotive art.

The early introduction of steam navigation into the country, and the rapid increase which has since taken place in the number of steam-boats, have afforded an extensive field for the prosecution of valuable inquiries on this interesting subject; and the builders of steam-boats, by availing themselves of the opportunities held out to them, have been enabled to make constant accessions to their practical knowledge, which have gradually produced important improvements in the construction and action of their vessels. But on minutely examining the most approved American steamers, I found it impossible to trace any *general* principles which seem to have served as guides for their construction. Every American steam-boat builder holds opinions of his own, which are generally founded, not on theoretical principles, but on deductions drawn from a close examination of the practical effects of the different arrangements and proportions adopted in the construction of different steam-boats, and these opinions never fail to influence, in a greater or less degree, the built of his vessel, and the proportions which her several parts are made to bear to each other.

So lately as twelve years ago, about thirty hours were occupied by the steam-boats navigating the Hudson in making their passages from New York to Albany, a distance of about one hundred and fifty miles, which is at the rate of only five miles per hour. Passengers were then conveyed in barges towed by steam-boats, to avoid the danger which, according to the following extract from an advertisement of the sailing of the vessels, seems at that time to have attended the steam navigation of the country: "Passengers on board the safety barges will not be in the least exposed to any accident which may happen by reason of the fire or steam on board of the steam-boats. The noise of the machinery, the trembling of the boat, the heat from the furnace, boilers, and kitchen, and every thing which may be considered as unpleasant or dangerous on board of a steam-boat, are entirely avoided." These "safety barges," however, have been entirely laid aside, and the voyage between Albany and New York is now generally performed in ten hours, exclusive of the time lost in making stoppages, being at the astonishing rate of fifteen miles per hour. They have effected this great increase of speed by constantly making experiments on the form and proportions of their engines and vessels, in short, by a persevering system of *trial and error*, which is still going forward; and the natural consequence is, that, even at this day, no two steam-boats are alike, and few of them have attained the age of six months without undergoing some material alterations.

These observations apply more particularly to the steamers navigating the Eastern Waters of the United States, where the great number of steam-boat builders, and the rapid increase of trade, have produced a competition which has led to the construction of a class of vessels unequalled in point of speed by those of any other quarter of the globe. The original construction of most of these vessels has, as already stated, been materially changed. The breadth of beam and the length of keel have in some vessels been increased, and in others they have been diminished. This mode of pro-

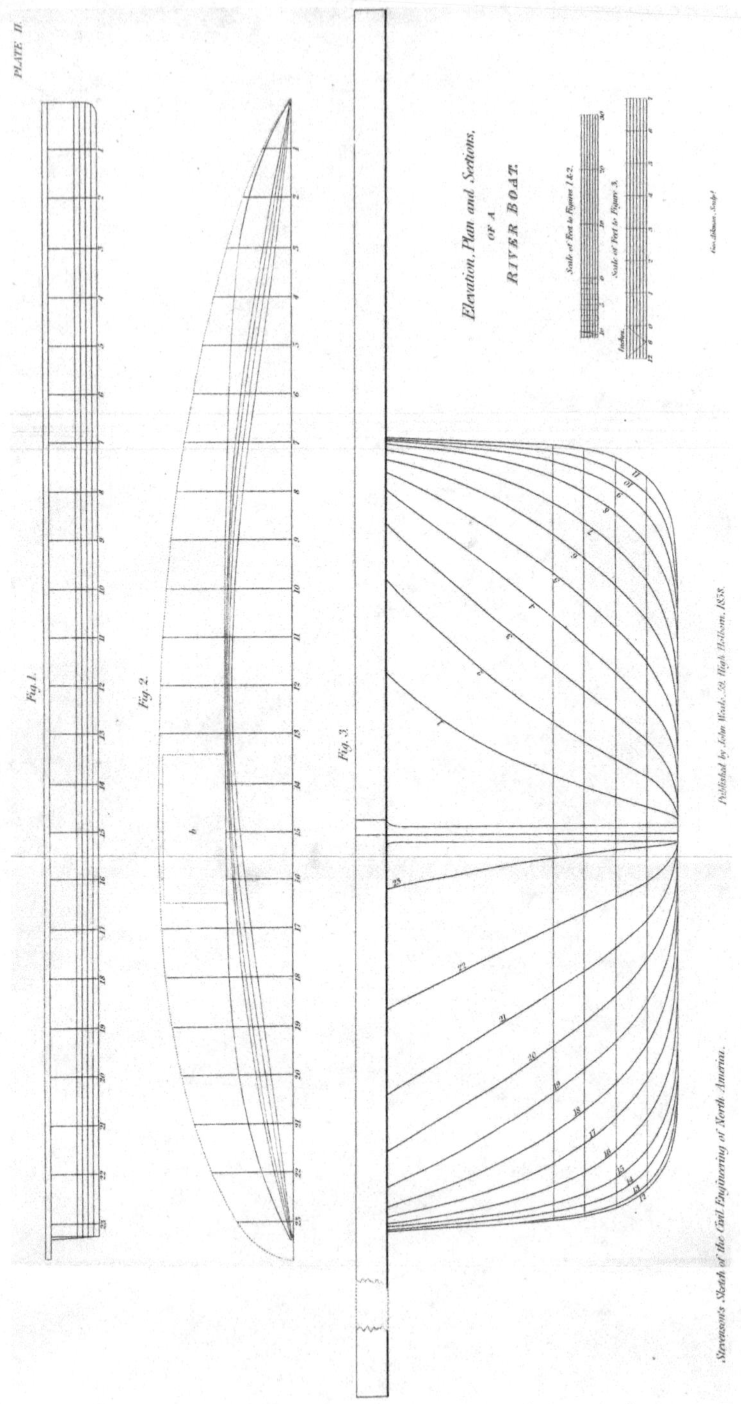

The material originally positioned here is too large for reproduction in this reissue. A PDF can be downloaded from the web address given on page iv of this book, by clicking on 'Resources Available'.

cedure may seem rather paradoxical ; but in America it is no uncommon thing to alter steam-boats by cutting them through the middle, and either increasing or diminishing their dimensions as the occasion may require. It is only a short time since many of the steam-boats were furnished with false bows, by which the length of the deck and the rake of the cutwaters were greatly increased. On some vessels these bows still remain ; from others they have been removed, subsequent experiments having led to the conclusion, that a perpendicular bow without any rake, as shewn in Plate II. fig. 1, is best adapted for a fast sailing boat. When I visited the United States in 1837, the "Swallow" held the reputation of being one of the two swiftest steamers which have ever navigated the American waters, and this vessel had received an addition of twenty-four feet to her original length, besides having been otherwise considerably changed. Before these alterations were made on her, she was considered, as regards speed, to be an inferior vessel.

The inferences to be drawn from these facts are, that the great experiment for the improvement of steam navigation, in which the Americans may be said to have been engaged for the last thirty years, is not completed, and the speed at which they have succeeded in propelling their steam-vessels may yet be increased ; and also that, in the construction of their vessels, they have been governed by experience and practice alone, without attempting to introduce theoretical principles, in the application of which, to the practice of propelling vessels, by the action of paddle-wheels on the water, numerous difficulties have hitherto been experienced.

There are local circumstances, connected with the nature of the trade in which the steam-boats are engaged, and the waters which they are intended to navigate, that have given rise to the employment of three distinct classes of vessels in American steam navigation, all of which I had an opportunity of sailing in and particularly examining.

These steam-boats may be ranged under the following classification: First, those navigating the Eastern Waters. This class includes all the vessels plying on the River Hudson, Long Island Sound, Chesapeake and Delaware Bays, and all those which run to and from Boston, New York, Philadelphia, Baltimore, Charleston, Norfolk and the other ports on the eastern coast of the country, or what the Americans call the Sea-board. Second, those navigating the Western Waters, including all the steamers employed on the river Mississippi and its numerous tributaries, including the Missouri and Ohio. Third, the steamers engaged in the Lake navigation. These classes of vessels vary very much in their construction, which has been modified to suit the respective services for which they are intended.

The general characteristics by which the Eastern Water boats are distinguished, are, a small draught of water, great speed and the use of condensing engines of large dimensions, having a great length of stroke. On the Western Waters, on the other hand, the vessels have a greater draught of water and less speed, and are propelled by high-pressure engines of small size, worked by steam of great elasticity. The steamers on the Lakes, again, have a very strong built and a large draught of water, possessing in a greater degree the character of *sea*-boats than any of those belonging to the other two classes. They also differ in having masts and sails, with which the others are not provided.

The steam-boats employed on the Hudson River are the first, belonging to the class of vessels navigating the Eastern Waters, of which I shall make particular mention.

The shoals in the upper part of the river, produced by the Overslaugh which I formerly mentioned, have rendered it necessary that the steam-boats employed in its navigation should have a small draught of water. The great trade of the river, and the crowds of passengers which are constantly travelling between New York and Albany and the intermediate towns, have also led to the adoption of separate lines of boats, one for towing barges loaded with goods, and another devoted exclusively to the conveyance of passengers. The attainment of great speed naturally became an important desideratum in the construction of the vessels employed in carrying passengers; and the success which has attended the efforts of the steam-boat builders to produce vessels, combining swiftness with efficiency and perfection of workmanship, is truly wonderful, and in the highest degree creditable.

A table will be found at page 169, containing the dimensions of several of the steam-boats running in America, which I had an opportunity of examining when I visited the country in 1837. Among these the dimensions of several of the Hudson boats are given; but in order to explain more clearly the general arrangement of their parts and mode of operation, I shall give in detail the dimensions of the steam-boat "Rochester," plying between New York and Albany. The elevation, plan, and water-lines of this vessel are shewn in Plate II.* The most satisfactory observations which I was able to make relative to the maximum speed at which the American steam-boats are capable of being propelled, were made during a passage in the "Rochester," which serves as a further motive for particularly describing her construction.

The "Rochester" measures 209 feet ten inches in length on her deck. This measurement applies also to the length of her keel, her stern-post and cut-water being perpendicular, as shewn in Plate II. The maximum breadth of beam is 24 feet. The projection of that part of the deck called the wheel-guards,

* The lines of the steamers in Plates II. and III. were laid down by my friend Mr Andrew Murray, of Messrs Fairbairne and Murray, from models which I brought from New York.

shewn in dotted lines in Fig. 2, beyond the hull of the vessel, is 13 feet on each side. The maximum breadth of the vessel, measured to the outside of the paddle-wheels, is 47 feet. The depth of hold is 8 feet 6 inches. The draught of water, with an average number of passengers, is four feet. The diameter of the paddle-wheels is 24 feet. The length of the float-boards, which are twenty-four in number, is 10 feet. The dip of the float-boards is 2 feet 6 inches. This vessel is propelled by one engine, having a cylinder of 43 inches in diameter, and the length of stroke 10 feet. The engine condenses the steam which works expansively, and is cut off at half stroke.

The great competition that exists in the navigation of the Hudson produces constant *racing* between boats belonging to different companies; and this is not unfrequently attended with serious accidents. When the "Rochester" is pitched against another vessel, and at her full speed, the steam is often carried as high as forty-five pounds on the square inch of the boiler; and the piston makes twenty-seven double strokes, or, in other words, moves through a space of 540 feet per minute, or 6.13 miles per hour. In this case the circumference of the paddle-wheels moves at the rate of 23.13 miles per hour. In ordinary circumstances, however, the engine is worked by steam of from twenty-five to thirty pounds pressure on the square inch; and in this case the piston makes about twenty-five double strokes per minute, moving through a space of 500 feet per minute, or 5.68 miles per hour; and the circumference of the paddle-wheel moves at the rate of 21.42 miles per hour. The rate at which the pistons of marine engines in this country move, seldom exceeds 210 feet per minute. The pistons of locomotive engines generally move at the rate of about 300 feet per minute; but both of their speeds are very far short of the velocity of the "Rochester's" piston.

The hulls of almost all the American steam-boats, especially those which ply on the rivers, carrying no freight excepting the luggage belonging to passengers, are constructed in a very light and superficial manner. They are built perfectly flat in the bottom, and perpendicular in the sides; a cross section in the middle of the vessel, having the form of a parallelogram, with its lower corners rounded off, as shewn by the cross sections in Plate II. This construction of hull is well adapted to a navigation where the depth of water is small, and the attainment of great speed is an object of importance, as it insures a smaller draught of water, and consequently affords less resistance to the motion of the vessel than any other mould which has an equal area of cross section below the water line; but vessels built in this way, without a deep keel, having no hold of the water, are not well adapted for making sea-voyages, as they cannot resist the effect of the wind, which causes them to make lee-way. It is only the great breadth of the paddle-wheels and power of the engines which enables the American boats to move steadily through the water. The breadth of the paddle-wheels is, in fact, so much additional breadth added to the beam of the vessel; for the reaction of the float-boards striking the water tends, in some measure, to counteract any tendency that the vessel may have to roll, which would otherwise be very apt to take place in the American steamers, where the machinery and boilers are placed above the level of the deck. There is no rolling motion felt in these fast boats. The rectilineal motion, however, is by no means regular. Every stroke of the engine produces a momentary acceleration in the speed, giving rise to a *see-saw* motion, resembling that of a row-boat, in which the impulse produced by every stroke of the oars is distinctly felt.

In the American steamers the keel generally projects from two to six inches from the bottom of the hull, and is level from stem to stern. Its principal service, when the projection is so small, consists in strengthening the hull. The deck-lines of the hull, in general, begin to fall in at a distance of a few feet from the middle of the vessel. They approach each other with a gentle curve, as shewn in Plate II. Fig. 2, towards the stern and bow, where they meet, and are connected by the stern-post and cutwater of the vessel. The cutwater is generally perpendicular, and the sides of the vessel, diverging from it, present a very acute angle to meet the resistance offered by the water. The angle formed by the sides of the "Rochester" is about twenty degrees at the level of the water, and decreases to about ten degrees at the level of the keel.

The engine and paddle-wheels are placed in a frame-work of wood, to which they are attached by strong fixtures. This frame-work is generally a specimen of substantial and excellent workmanship. The timbers of which it is composed are arranged so as to form the frustum of a pyramid. The apex of the framing is elevated above the deck and paddle-wheels, and supports the walking-beam of the engine, while its base rests on the flooring timbers of the hull. In this way the weight of the machinery is distributed over a large surface of the bottom, the weak construction of that part of the vessel rendering such an arrangement absolutely indispensable to her safety. Iron rods, fastened to the timbers of the vessel, extend fore and aft from the upper part of the beams forming the engine framing. These iron ties give support to the bow and stern, which invariably sink or settle down in the course of a few months, owing to the slim built and great length of the hull, if not braced up in the manner described. Screws and nuts are generally provided, by which the ties can be tightened up, should any yielding take place in the wood-work of the vessel.

At the height of about five feet above the surface of the water the hull is covered with a deck. It is

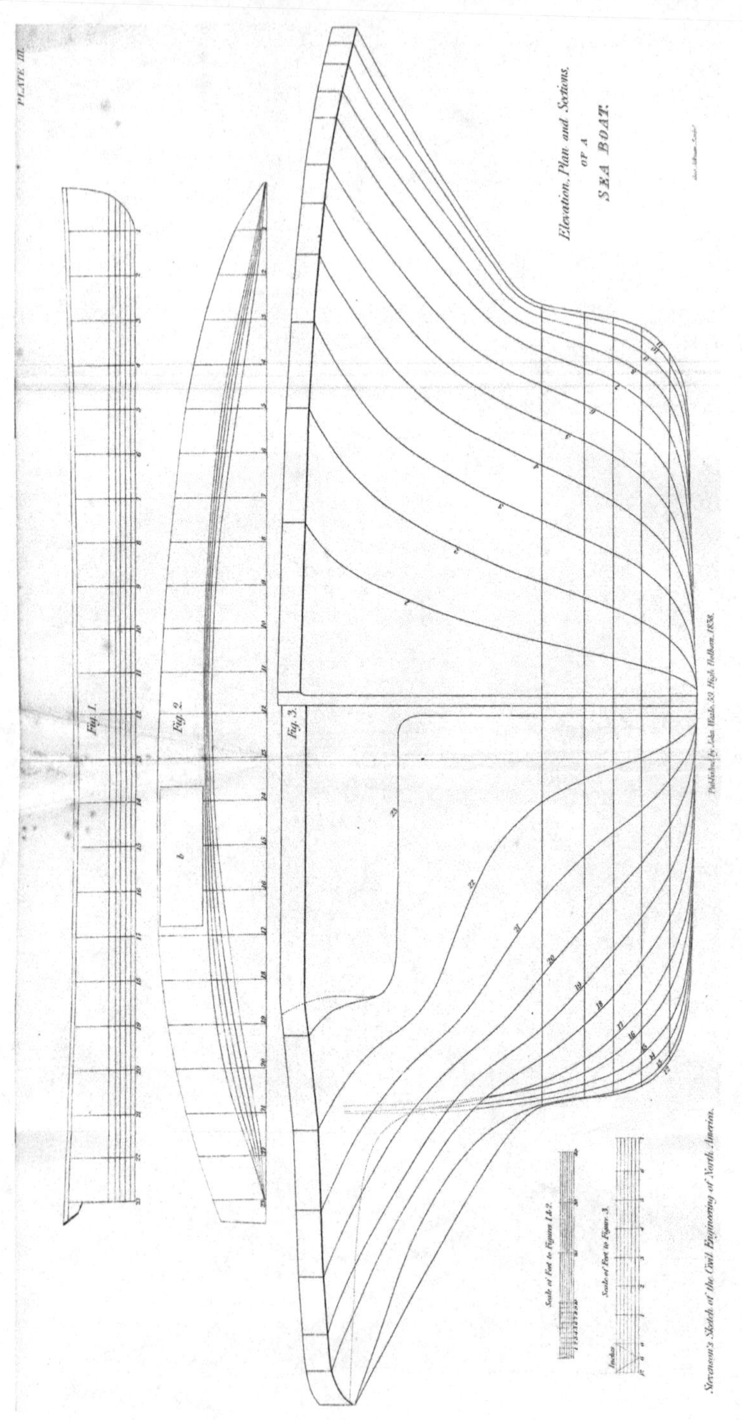

The material originally positioned here is too large for reproduction in this reissue. A PDF can be downloaded from the web address given on page iv of this book, by clicking on 'Resources Available'.

generally made somewhat in the form of an ellipse, as shewn by the dotted lines in Plate II. Its vertices rest on the stern-post and cutwater of the vessel, while its sides, expanding beyond the hull, overhang the water, and the bulwarks of the vessel are erected on its circumference. The part of the deck overhanging the water is called the wheel-guards, and in some vessels it has a projection of 18 or 20 feet from the sides. In the " Rochester," the projection, as I have already said, is 13 feet. The wheel-guards are formed so as to inclose the paddle-wheels, which work in spaces left in them for that purpose, marked *b* in Plates II. and III. The inner plumber-blocks and paddle-wheel axles rest on the timbers of the vessel, and the exterior ones on the outer edges of the guards.

A large cabin, serving the double purpose of the gentlemen's sleeping apartment and the public dining-room, is formed in the hull of the vessel. It is entered by a stair leading from the first deck. It generally extends nearly from stem to stern, and is elegantly fitted up. The ladies' cabin is on a level with the first deck, from which it enters. This deck is covered with a roof extending from the paddle-wheels to the stern of the vessel, the top of which forms a higher deck, raised about sixteen feet above the level of the water, called the promenade-deck. The general arrangement of these vessels will be best understood by referring to Plate IV., which is a perspective view of the steam-boat " Swallow.'

The vessels propelled by two engines carry two boilers and four funnels, and have a very extraordinary appearance. The vessels of modern construction, however, have generally only one engine, with two boilers and two smoke tubes, as shewn in the Plate of the "Swallow." The boilers are on a level with the lower deck, and rest on the wheel-guards, one being placed on each side of the vessel. The cylinder, which also stands on a level with the first deck, is placed in the centre of the vessel, between the two boilers. The condenser and pumps are situate in the hull of the vessel, in the middle of the large cabin, from which, they are separated by a wooden partition.

Engines working with side-rods, connected by a cross-head, which is attached to the end of the piston-rod, and moves in vertical slides, are occasionally employed in the steam-boats which navigate the Eastern Waters. The beam-engine is, however, much more generally used. The length of stroke adopted by the Americans for their marine engines, is very much greater than I have ever found in Europe. This renders it necessary that the main centres of the engine, or the pivots on which the beam performs its motion, should be placed at a considerable elevation above the promenade deck. The walking beam, therefore, is quite exposed, and is elevated above every other part of the vessel, excepting the tops of the smoke-tubes, as is shewn in Plate IV., forming one of the most prominent and striking parts of an American

STEAM BOAT SWALLOW, PLYING ON THE RIVER HUDSON. PLATE IV.

James Andrews, Delt. Geo Aikman, Sculpt.

Stevenson's Sketch of the Civil Engineering of North America.

Published by John Weale, 59, High Holborn, 1838.

steam-boat, and presenting, as may naturally be supposed, a strange effect in the eyes of those accustomed to see European steam-boats only, in which no part of the machinery is visible even from the deck of the vessel. The beams are constructed wholly of malleable iron, in the manner shewn in the following diagram—

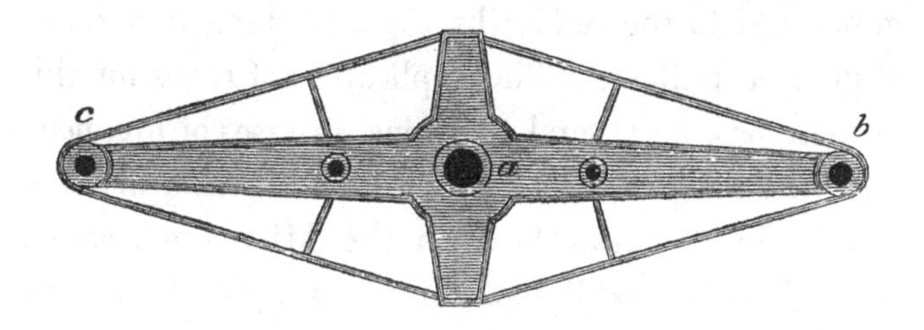

in which *a* is the main centre, and *b* and *c* the points to which the piston and connecting rods are attached. This construction combines lightness with strength and rigidity, and is found to act very well.

The arrangement of the decks and machinery which I have just described, and which is represented in Plates IV. and V., renders the vessel's course, when she is under weigh, quite invisible from her stern, and, consequently, it is impossible to steer her from that part of the ship; but the wheel by which the rudder is moved is placed in a wheel-house, erected for the pilot on the fore part of the promenade-deck, and in some instances at a distance of nearly 200 feet from the stern of the boat. The steersman, by this arrangement, stands so far forward in the vessel, and in so elevated a situation, that he cannot easily discover when the vessel swerves from her course, without the assistance of a tall perpendicular pole, placed at the bow, in the manner shewn in the plates. On this he keeps his eye, and, by narrowly observing its position in relation to some fixed object at a distance, he readily detects the smallest deviation from the course.

The motion produced by moving the wheel is communicated to the rudder by ropes working in a series of grooved pulleys. The application of ropes for this purpose has, on several occasions, in cases of fire, been attended with most unhappy results. During my stay in America, a steam-boat on the Mississippi, called the "Ben Sherod," took fire, and upwards of one hundred lives were lost, in consequence of the vessel's becoming unmanageable owing to the rudder ropes being burned. Iron rods and chains have lately been introduced instead of ropes, and will, doubtless, soon come into general use.

The rudder in general measures about 6 feet in depth, and 8 feet in length. It moves on pivots, which work in gudgeons fixed to the stern of the vessel, and thus far resembles the rudder used in all sea-vessels. The ropes, however, by which it is put in motion are made fast to the outer extremity of the rudder, in the manner shewn in the annexed diagram; and in this way the tiller, which takes up much room, is altogether dispensed with.

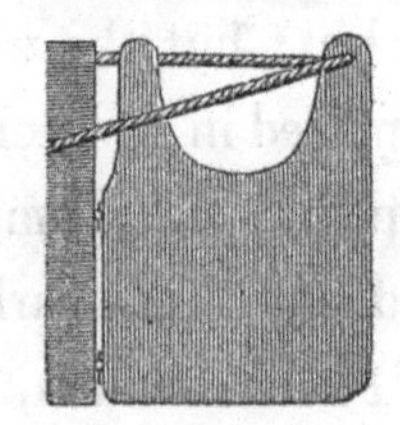

This mode of steering in an elevated situation, near the bow of the vessel, is peculiarly well adapted for steamers navigating narrow rivers, such as the Thames and Clyde in this country, which are crowded with craft of all kinds. On the suggestion of Captain Basil Hall, it has been introduced, a short time ago, on the Thames, in the steamer "Adelaide." It is singular that it is not in general use on such a river as the Thames, on which serious accidents, from the collision of vessels, are of so frequent occurrence, and where it is utterly impossible that a steersman, placed at the stern, can direct the vessel properly.

The foregoing remarks regarding the construction of the steamers refer particularly to those vessels navigating the rivers on the eastern coast of the United States. Those used on the bays and sounds, called sea-boats by the Americans, are somewhat different in their construction, their hulls and machinery being more strongly made, and their draught of water considerably greater. The river-boats draw from four to six feet of water, and the sea-boats from five feet six inches to nine feet; but still the machinery and boilers, as well as a great part of the cabin-accommodation in that class of steamers, is elevated above the level of the deck; an arrangement which seems very ill adapted for vessels exposed to the heavy gales and rough seas of the ocean. The best specimens of the American sea-boats are those which ply between New York and the ports of Providence and Charleston.

The finest of these sea-boats, and indeed the finest steamer which I saw in the United States, is the "Narragansett," plying between New York and Providence, which is shewn in Plate III. Fig. 1. is an elevation of the hull; Fig. 2. a plan; and Fig. 3. shews her water-lines. It could hardly be credited, from a mere examination of the drawings, that this vessel plies regularly from New York to Providence. By inspecting the map, it will be seen that, during the fifty miles of this voyage, extending between New London and Newport, she is quite exposed to the roll of the Atlantic Ocean; and, notwithstanding this, she makes her passages with great speed and regularity.

The "Narragansett" measures 210 feet in length of keel, and 26 feet in maximum breadth of beam. The depth of her hold is 10 feet 7 inches, and her draught of water is 4 feet 6 inches without the keel, and 5 feet with the keel, when she has her average load on board. She is built entirely of oak, and is strengthened by diagonal straps or ties of iron which connect her timbers. The vessel is propelled by one condensing engine, which works expansively, cutting off the steam at half stroke. The condensation of the steam in this engine, as well as in most of the American marine engines, is produced by the injection of a jet of cold water into the condenser. She carries two boilers, in which an aggregate amount of 3000 square feet of surface is exposed to the fire, and works with steam of a pressure varying, according to circumstances, from twenty to twenty-five pounds on the square inch. The cylinder is placed horizontally, and is 56 inches in diameter; the length of the stroke is 11 feet 6 inches, and the piston makes twenty-four double strokes per minute, so that its average motion in the cylinder is at the rate of no less than 6.27 miles per hour. The diameter of the paddle-wheels is 25 feet, and, as they perform twenty-four revolutions in the minute, the motion of the periphery is at the rate of 21.4 miles per hour. The breadth of the "Narragansett's" paddle-wheels is 11 feet, and their dip 2 feet 2 inches. The diameter of the paddle-wheel axle on which they are keyed is 13 inches.

The cabins of the sea-steamers are of great size, and their accommodation for passengers is excellent. In most of them about four hundred berths are provided. The principal cabin in the "Massachusetts," a vessel running on the line between New York and Providence, is 160 feet in length, about 22 feet in maximum breadth, and 12 feet in height; and, what adds greatly to its convenience and capacity, it is entirely unbroken by pillars or any other obstruction throughout its whole area. I have dined with 175 persons in this cabin; and, notwithstanding this numerous assembly, the tables, which were arranged in two parallel rows extending from one end of the cabin to the other, were far from being fully occupied; the attendance was good, and every thing was conducted with perfect regularity and order. There are 112 fixed berths ranged round this cabin, and about 100 temporary berths can be erected in the middle of the floor. Besides these, there are 60 fixed berths in the ladies' cabin, and several temporary sleeping-places can be erected in it also. The cabin of the "Massachusetts" is by no means the largest in the United States; some steamers have cabins upwards of 175 feet in length. Those large saloons are lighted by argand lamps suspended from the ceiling, and their appearance, when brilliantly lighted up and filled with company, is very remarkable. The passengers generally arrange themselves in parties at the numerous small tables (into which the large tables are converted after dinner), and engage in different amusements. The scene resembles much more the coffee-room of some great hotel than the cabin of a floating vessel.

I found no variety in the construction of the paddle-wheels of the different American steam-boats. They are all made in the manner represented in the following diagram. The spokes are made of wood, and

bolted into cast-iron flauges which are keyed to the axle of the paddle-wheel; their outer ends are connected together by bands of iron encircling the circumference of the wheel. The float-boards, which are formed of hardwood, are attached to the spokes simply by bolts. The float-boards do not extend across the whole breadth of the paddle-wheel, as is always the case in this country. They are divided into two and sometimes three compartments, and the wheel is furnished with three and sometimes four sets of spokes arranged in parallel planes. "This construction was introduced by Mr Stevens of New York, and may be described," says Dr Renwick, "by supposing a common paddle-wheel to be sawn into three parts in planes perpendicular to its axis. Each of the two additional wheels that are thus formed, is then moved back, until their paddles divide the interval of the paddles on the original wheel into three equal parts.

"In this form the shock of each paddle is diminished to one-third of what it is in the usual shape of the wheel; they are separated by less intervals of time, and hence approach more nearly to a constant resistance; while each paddle following the wake of those belonging to its own system strikes upon water that has been but little disturbed."*

The large diameter of the American paddle-wheels renders unnecessary the use of the cycloidal paddle of

* Treatise on the Steam-Engine by James Renwick, LL.D., New York 1830.

Mr Galloway, or the eccentric paddle of Mr Morgan, now frequently adopted in this country to obviate the evils arising from indirect impulse and backwater, which affect so powerfully the action of paddle-wheels of small diameter. In some of the Western Water boats, which are often very deeply laden, the paddle-wheels are constructed with moveable float-boards, so that their dip may be increased or diminished to suit the draught of water; but this construction, so far as I know, is not in use in any other part of the country.

The American steamers are generally propelled only by one engine, and a counter-balance attached to the paddle-wheels is in some cases found necessary, to enable the engine to turn its centres. The great length of the stroke, however, allows time for a degree of momentum to be generated, which is sufficient in most cases to carry the engine past its centres, and failing this, the paddle-wheels, from their large diameter, become good generators of momentum, and act in the same way as the fly-wheels of land engines in regulating their motion. Even in those vessels where two engines are employed, their connecting-rods are not attached to the same axle; each engine works quite independently of the other, and drives only one of the paddle-wheels; whereas in this country the connecting-rods of both engines are attached to the same axle, by cranks placed at right angles to each other, so that one engine is exerting its full power at the very moment when the other is expending none of its force, and the power is thus employed in the most advantageous manner for keeping up the speed. The short stroke and comparatively small diameter of the paddle-wheels in European boats, renders this construction necessary to enable engines to pass their centres.

The general construction of the boilers, and the arrangement of the flues, in the steam-boats on the Eastern Waters, resemble in a great measure those of European steamers. The flame and smoke generated in the fire-place by the combustion of the fuel, pass through flues in the interior of the boiler, and are afterwards discharged into the smoke-tube. The boilers are strengthened in the usual manner, by means of iron braces or ties, arranged so as to form a strong connection between the interior surfaces, and thus render them more capable of resisting the expansive force of the steam, which has a tendency to tear them asunder. Copper was, until lately, very generally employed in America for the construction of the boilers of vessels navigating the sea, this metal being less liable than iron to be acted on by the saline deposits. By means of some improvements which have lately been introduced, these deposits are prevented from collecting in iron boilers to any dangerous extent, and the difference of expense is so much in favour of iron, that it has now been adopted instead of copper, in the sea, as well as in the river boats. The means used in America for checking the deposit which takes place in boilers from the use of salt-water, is the same as that generally employed in this country, namely, by "Blowing off," an operation which is performed every two or three hours, while the boat is running, without stopping her progress. A valve in the bottom of the boiler being opened, part of the water is permitted to escape, which, in its rush from the boiler, disturbs any deposit that may have taken place on its bottom, and generally carries it off.

The speed of the American steam-boats has excited considerable wonder in this country ; and some people have been inclined to doubt the accuracy of the statements that have frequently been made regarding the extraordinary feats performed by them. Fast sailing is a property which is not possessed by all American steam-boats ; but that a few of those navigating the River Hudson and Long Island Sound perform their voyages safely and regularly, at a speed which far surpasses that of any European steamer hitherto built, every impartial person, who has had an opportunity of seeing the performances of the vessels in both countries, must be ready to admit.

Some difficulties at present exist, which preclude the attainment of more than an approximation in ascertaining the maximum rate at which the steam-boats on the Hudson are capable of being propelled in still water. One of these is caused by the currents of the flowing and ebbing tide, which are felt as far as Albany, and whose velocity has never been accurately ascertained, and the other by the doubt that exists as to the actual distance of the route between New York and Albany, which has been variously stated at from 145 to 160 miles. The road between these towns runs nearly parallel to the river, and is said to be 162 miles in length. In the American Almanac for 1837, the town-house of New York is stated to be in north latitude 40° 42′ 40″, and west longitude (from Greenwich) 74° 1′ 8″, and that of Albany in north latitude 42° 39′ 3″, and west longitude 73° 44′ 49″, which makes the distance between the two places, as the crow flies, 134.5 statute miles. The navigable channel of the Hudson, however, is by no means straight; its direction ranges over fifteen points of the compass, from West to E.N.E., including an angle of 157° 30′. Mr Redfield of New York, who has bestowed much attention on the subject of steam navigation, is of opinion that the length of the steam-boat route is 150 miles, being 15.5 miles greater than the distance measured by a straight line drawn between the two places.* This may be regarded as a near approximation to the truth. The same difficulties occur regarding the length of the routes performed by the boats navigating Long Island Sound, and the strength of the tidal currents encountered by them. It is quite evident that until these facts are accurately ascertained, it is impossible, with

* Professor Silliman's Journal, vol. xxiii. p. 312.

out a series of experiments made solely with that object in view, to discover what is the actual speed generally attained by American steam-boats. A very general opinion exists in America on this subject, in which many persons possessing the best means of information concur, that the fast steam-boats in that country can be propelled at the rate of eighteen miles an hour in still water, a feat which it is said has of late been often performed. I cannot vouch for the accuracy of this statement, however, from personal experience or observation; but this I can state positively, that the average length of time occupied by the steamers in making the voyage from New York to Albany, is ten hours, exclusive of time lost in making stoppages, which, taking the distance at 150 miles, gives fifteen miles an hour as their average rate of motion.

The "Rochester" and the "Swallow" were said to be the two swiftest boats running on the Hudson in 1837. I made a trip from Albany to New York in the "Rochester," on the 14th of June, on which occasion, with a view to test the vessel's speed, I carefully noted the hour of departure from Albany, the times of touching at the several towns and landing places on the river, with the reputed distances between them, the number of minutes lost at each place, and the hour of arrival at New York. Thirteen stoppages, which I found to average three minutes each, were made to land and take on board passengers. The

"Rochester" performed the voyage in ten hours and forty minutes. From this, thirty-nine minutes must be deducted for the time lost in making the thirteen stoppages, which leaves ten hours and one minute as the time during which the vessel was actually occupied in running from Albany to New York Assuming the distance between those places to be 150 miles, the average speed of the vessel throughout the trip was 14.97 miles per hour, but even if we assume the distance to be only 145 miles (the shortest distance I have ever heard stated), which there is every reason to believe is too small, the average rate is still 14.47 miles per hour, the difference of five miles in the length of the route, producing a diminution in the vessel's average rate of sailing of but half a mile per hour. The current was in the "Rochester's" favour during the first part of the voyage, but the Overslaugh shoals, and the contracted and narrow state of the navigable channel of the river for about thirty miles below Albany, checked her progress very much; and, consequently, for the first twenty-seven miles her speed was only 12.36 miles per hour. This was her average rate of sailing during the part of her course when her speed was slowest. After the first thirty miles the river expanded, affording a better navigable channel, when her speed gradually increased, and before the flowing tide checked her progress the vessel attained the maximum velocity indicated by my observations, which, between two of the stopping places, was 16.55 miles per hour. When going at this speed it is possible that she was influenced by some slight degree of current in her favour, although it was quite imperceptible to the eye, as the flow of the tide appeared to produce a stagnation in the water of the river. At West Point we encountered the flood tide, as was very distinctly proved by the swinging of the vessels which lay at anchor in the river. After this we had an adverse current all the way to New York, a distance of about fifty miles, and the vessel's speed during this part of the voyage averaged 14.22 miles an hour. About one half of the voyage was thus performed with a favourable current, and the other half was performed under unfavourable circumstances, owing partly to the shallowness of the water and the narrowness of the channel in the upper part of the river, and partly to an adverse tide in the lower part of it. When the Rochester is pitched against another vessel and going at her full speed, her piston, as formerly stated, makes twenty-seven double strokes per minute. On the voyage above alluded to, however, the piston, on an average, made about twenty-five double strokes per minute, so that the speed of 14.97 miles per hour, which she attained on that occasion, cannot be taken as her greatest ordinary rate of sailing. During the time, however, at which her speed was 16.55 miles per hour, her piston was making twenty-seven double strokes per minute, and at that time the vessel could not be far from having attained the maximum speed at which her engines are capable of propelling her through the water.

The rate of sixteen and a half miles an hour is very great, but perhaps not more than is due to the form of the vessels, and the power of the engines by which they are propelled. The "Rochester" draws only four feet of water, but the power of her engine is greater than that of any steamer in this country. The construction of the American marine engines is so different from that adopted in Europe, that it is doubtful if the same rule for calculating the power is applicable in both cases. In the following calculations, the deductions for the friction and for the difference between the pressure exerted by the steam in the boiler and in the cylinder, as well as the advantage that is derived from the use of a condenser, are in accordance with what has been stated by American engineers, who are best able to judge of the power of their own engines.* The diameter of the Rochester's piston is 43 inches, and its area is 1452.2 square inches. The pressure of the steam in the boiler is 45 lb. on the square inch; and the engine works expansively, and cuts off the steam at half stroke. The half of that pressure, or 22.5 lb., is assumed as the pressure acting on the square inch of the piston. To this, 10 lb. is added as the pressure of the atmosphere obtained by the use of the condenser, making the whole effective pressure on every square inch of

* Professor Silliman's Journal, vol. xxiii. p. 315.

the piston's area 32.5 lb. The length of the stroke is 10 feet, and, when going at full speed, the piston makes 27 double strokes, or, in other words, moves through the space of 540 feet every minute. Estimating the power of a horse as equal to that exerted in raising 33,000 lb. 1 foot per minute, the power of the engine is obtained by the following expression:

$$\frac{1452.2 \times 32.5 \times 540}{33000} = \frac{25486110}{33000} = 772.3$$

From this it appears, that a force is exerted upon the engine equal to that of 772.3 horses; but one-third of this power is supposed to be expended in working the pumps and overcoming the friction of the machinery, and a power of 514.8 horses remains as the true force exerted in propelling the vessel. The "Narragansett," as formerly noticed, draws five feet of water, and the power of her engine, calculated on the same principles, and with the same deductions, is equal to that of 618 horses. If the calculation generally adopted in this country were applied to those engines, and only one-fourth of the power deducted, which appears to be an ample allowance for engines of that construction, the power of the "Rochester" would be equal to 748, and that of the "Narragansett" to no less than 772 horses. The power of the "Great Western," plying between Bristol and New York, which is the largest steamer in this country, is said to be equal to that of 450 horses.

The disturbance created by the passage of the fast

American steamers through the water, is exceedingly small. The water, at the distance of twelve inches in front of their bows, presents a perfectly smooth and untroubled surface. A thin sheet of spray, composed of small globules of water, from a sixteenth to an eighteenth of an inch in diameter, rises nearly perpendicularly in front of the cut-water to the height of three, and, in some cases which I have observed as much as four feet, and falls again into the water on each side of the vessel. There is little or no commotion at the stern ; and the diverging waves which invariably follow the steamers in this country, and break on the banks of our rivers with considerable violence, are not produced by the fast boats in America. The waves in their wake are very slight, and, as far as I could judge, seem to be nearly parallel ; and the marks of the vessel's course cannot be traced to any great distance. These facts are quite in accordance with the result of some of Mr Russell's experiments, by which he was led to conclude that "the commotion produced in a fluid by a vessel moving through it, is much greater at velocities less than the velocity of the wave" (which is proportioned to the depth of the water), " than at velocities which are greater than it."*

Steam-boats were first introduced on the Mississippi in the year 1811, and in 1831 no less than 348 steam-

* Researches on Hydrodynamics, from the Transactions of the Royal Society of Edinburgh for 1837. By John Scott Russell, Esq.

ers had been built for the Western Water navigation, 198 of which were then in actual operation. Since that time their number has rapidly increased, with the increasing population and trade of the country, and is now said to be between 350 and 400; but, so far as I know, no official statement regarding the Western Water navigation has appeared since the publication of the following table, which is taken from the American Almanac for 1832, and contains a list of steamers up to that date, specifying those which have been worn out and have been lost to the service.

WHOLE NUMBER OF STEAM-BOATS BUILT ON THE WESTERN WATERS.

When built.	Whole Number.	Now running.	Lost or worn out.
1811	1	...	1
1814	4	...	4
1815	3	...	3
1816	2	...	2
1817	9	...	9
1818	23	...	23
1819	27	...	27
1820	7	1	6
1821	6	1	5
1822	7	...	7
1823	13	1	12
1824	13	1	12
1825	31	19	12
1826	52	36	16
1827	25	19	6
1828	31	28	3
1829	53	53	...
1830	30	30	...
1831	9	9	...
	348	198	150

Of the boats now running,

68 were built at Cincinnati.		
68	...	Pittsburg.
2	...	Louisville.
12	...	New Albany.
7	...	Marietta.
2	...	Zanesville.
1	...	Fredericksburg.
1	...	Westport.
1	...	Silver Creek.
1	...	Brush Creek.
2	...	Wheeling.
1	...	Nashville.
2	...	Frankfort.
1	...	Smithland.
1	...	Economy.
6	...	Brownsville.
3	...	Portsmouth.
2	...	Steubenville.
2	...	Beaver.
1	...	St Louis.
3	...	New York.
1	...	Philadelphia.
10	...	(Not known where.)
198		

Of the whole number, 111 were built at Cincinnati, 68 of which were running in 1831.

Of the 150 lost or worn out, there were—

Worn out,	63
Lost by snags,	36
Burned,	14
Lost by collision,	3
By other accidents not ascertained,	34
Total,	150

Most of the vessels at present employed have been built on the banks of the Ohio, and a few at St Louis, on the upper part of the Mississippi, but, according to the above list, the building-yards which have produced the greatest number are those of Pittsburg and Cincinnati, on the Ohio. Pittsburg, although about

2000 miles from the Gulf of Mexico, is a place of great trade. Its population is 30,000 persons, a great part of whom are employed in the construction and management of steam-boats, and some idea may be formed of the extent of their trade, when I state, that I have counted no less than thirty-eight steam-boats moored opposite the town in the Monongahela, all of which were engaged in plying to and from the port.

The vast number of vessels on the Western Waters, the peculiarity of their construction, and the singular nature of the navigation in which they are employed, make them objects of considerable interest to the traveller. We must not expect to find, however, in that class of vessels, the same display of good workmanship, and the attainment of the high velocities, which characterise the vessels on the Eastern Waters. These qualifications may be very easily dispensed with, and the want of them is by no means the worst feature in the western navigation; but, what is of far more importance, too many of the vessels are decidedly unsafe; and, in addition to this, their management is intrusted to men whose recklessness of human life and property, is equalled only by their ignorance and want of civilization.

Economy would indeed seem to be the only object which the constructors of these boats have in view, and therefore, with the exception of the finery which the cabins generally display, little care is expended in their construction, and much of the workmanship connected with them is of a most superficial and insuffi-

WESTERN WATER STEAM BOAT.

PLATE V.

Drawn by James Andrews, from a sketch made on the River Ohio, by David Stevenson.

Geo. Aikman, Sculp.t

Published by John Weale, 59, High Holborn, 1838.

Stevenson's Sketch of the Civil Engineering of North America.

cient kind. When the crews of these frail fabrics, therefore, engage in brisk competition with other vessels, and urge the machinery to the utmost extent of its power, it is not to be wondered at that their exertions are often suddenly terminated by the vessel taking fire, and going to the bottom, or by an explosion of the steam-boilers. Such accidents are frequently attended with an appalling loss of life, and are of so common occurrence, that they generally excite little or no attention. During my stay in North America, a steamer called the "Ben Sherrod," as already mentioned, was burnt on the Mississippi, when 120 persons were reported to have lost their lives. I am happy in being able to add, that there is reason to believe that, in consequence of this accident, the Government of the United States have resolved to take some measures to insure the better regulation of this navigation, which has been too long neglected by them.

The vessels on the Western Waters vary from 100 to 700 tons burden, and are generally of a heavy built, to enable them to carry goods. They have a most singular appearance, and are no less remarkable as regards their machinery. Plate V. is a perspective view of one of them, taken from a sketch which I made on the Ohio. They are built flat in the bottom, and generally draw from six to eight feet of water. The hull is covered with a deck at the level of about five feet above the water, and below this deck is the hold, in which the heavy part of the cargo is carried. The whole of the machinery rests on the first deck; the engines being placed near the middle of the vessel, and the boilers under the two smoke chimneys, as shewn in the drawings. The fire-doors open towards the bow, and the bright glare of light thrown out by the wood fires, along with the puffing of the steam from the escapement pipe, produce a most singular effect at night, and serve the useful purpose of announcing the approach of the vessel when it is still at a great distance. The chief object in placing the boilers in the manner described, is to produce a strong draught in the fire-place. The other end of the lower deck, which is covered in, and occupied by the crew of the vessel and the deck passengers, generally presents a scene of filth and wretchedness that baffles all description. A stair-case leads from the front of the paddle-boxes on each side of the vessel, to an upper gallery about three feet in breadth. This surrounds the whole after-part of the vessel, and is the promenade of the inhabitants of the second deck. Several doors lead from the gallery into the great cabin, which extends from the funnels to within about thirty or forty feet of the stern of the vessel; the aftermost space is separated from the great cabin by a partition, and is occupied by the ladies. The large cabin contains the gentlemen's sleeping berths, and is also used as the dining-room. This part of the western steamers is often fitted up in a gorgeous style; the berths are large, and the numerous windows by which the cabin is surrounded give abundance of light, and, what is of great consequence in that scorching climate, admit a plentiful supply of fresh air.

From the gallery surrounding the chief cabin, two flights of steps lead to the hurricane deck, which, in many of the steamers, is at least thirty feet above the level of the water. The wheel-house, in which the steersman is placed, is erected on the forepart of this deck, and the motion is communicated to the helm by means of ropes or iron rods, in the manner already described in speaking of the Eastern steamers.

The first cabin of a Mississippi steam-boat is strangely contrasted with the scenes of wretchedness in the lower deck, and its splendour serves in some measure to distract the attention of its unthinking inmates from the dangers which lie below them. But no one who is at all acquainted with the steam-engine, can examine the machinery of one of those vessels, and the manner in which it is managed, without shuddering at the idea of the great risk to which all on board are at every moment exposed.

The Western Water steamers are propelled sometimes by one and sometimes by two engines. When two engines are used, the ends of the piston-rods work in slides, and the connecting-rods are both attached to cranks on the paddle-wheel axle, placed at right angles to each other, as is the case in most of the steamers in this country. When only one engine is used, which is more generally the case, a large fly-wheel, from ten to fifteen feet in diameter, is fixed on the paddle-wheel shaft, and serves to regulate the motion of the engine, and enable it to turn its centres. The cylinders are invariably placed horizontally, and the engines are always constructed on the high-pressure principle.

The engines are generally very small in proportion to the size of the vessel which they propel, and, to make up for their deficiency in volume, they are worked by steam of great elasticity. The "Rufus Putnam," for example, a pretty large vessel drawing six feet of water, which plies between Pittsburg on the Ohio and St Louis on the Mississippi, is propelled by a single engine having a cylinder 16 inches diameter, and 5 feet 6 inches in length of stroke, but this engine is worked by steam of a most dangerously great elasticity. The captain of the vessel informed me that, under ordinary circumstances, the safety-valves were loaded with a pressure equal to 138 lb. on the square inch of surface, but that the steam was occasionally raised as high as 150 lb. to enable the vessel to pass parts of the river in which there is a strong current; and he added, by way of consolation, that this amount of pressure was never exceeded except on extraordinary occasions! I made a short voyage on the Ohio in this vessel, but after receiving this information, I resolved to leave her on the first opportunity that presented itself.

The "St Louis," one of the newest boats on the

Mississippi, is 230 feet in length of deck, and 28 feet in breadth of beam. She draws 8 feet of water, and carries about 1000 tons. This vessel is propelled by two engines, with cylinders 30 inches in diameter, and 10 feet in length of stroke, worked by steam having a pressure of 100 lb. on the square inch of the boiler.

Explosions, as may naturally be supposed, are of very frequent occurrence; and, with a view to cure this evil, several attempts have, at different periods, been made to introduce low-pressure engines on the Western Waters, but the cheapness of high-pressure engines, and the great simplicity of their parts, which require comparatively little fine finishing and good fitting, certainly afford reasons for preferring them to low-pressure engines, in a part of the country where good workmen are scarce, and where the value of labour and materials is very great. It must also be recollected, that a condensing or low-pressure engine takes up a great deal more space than one constructed on the high-pressure principle. I do not apprehend, however, that the number of accidents would be diminished by the simple adoption of low-pressure boilers, without the strict enforcement of judicious regulations; and if those regulations were properly applied to high-pressure boilers, they would not fail to render them perhaps quite as safe as those boilers which are generally made for engines working on the low-pressure principle. One very obvious improvement on the present hazardous state of the Mississippi navigation, would be the enactment of a law that the pressure of the steam should in no case exceed perhaps 50 lb. on the square inch.

The boilers of these steamers are all tubular, and have circular flues in them, which permit the passage of the flame through the body of the boiler. Those of the St Louis are nine in number. They are 42 inches in diameter, and 24 feet in length. Two circular flues 16 inches in diameter pass through the interior. The whole of the flues and outer coating of the boiler are made of sheet-iron three-sixteenths of an inch in thickness, and the end plates are formed of materials of greater strength. The boiler is strengthened by numerous internal ties, and is calculated to sustain a pressure of 100 lb. on the square inch of surface. The only protection which the boilers have from the atmosphere is a layer of clay, with which they are in all cases covered to prevent the radiation of heat.

The steamers make many stoppages to take in goods and passengers, and also supplies of wood for fuel. The liberty which they take with their vessels on these occasions is somewhat amusing, and not a little hazardous. I had a good example of this on board of a large vessel called the "Ontario." She was sheered close inshore among stones and stumps of trees, where she lay for some hours taking in goods. The additional weight increased her draught of water, and caused her to heel a good deal, and when her engines were put in motion, she actually crawled into the deep water on her paddle-wheels. The steam had been got up to an enormous pressure to enable her to get off, and the volumes of steam discharged from the escapement pipe at every half stroke of the piston made a sharp sound almost like the discharge of fire-arms, while every timber in the vessel seemed to tremble, and the whole structure actually groaned under the shocks.

During these stoppages, it is necessary to keep up a proper supply of water to prevent explosion, and the manner in which this is effected on the Mississippi is very simple. The paddle-wheel axle is so constructed, that the portions of it projecting over the hull of the vessel to which the wheels are fixed can be thrown out of gear at pleasure by means of a clutch on each side of the vessel, which slides on the intermediate part of the axle, and is acted on by a lever. When the vessel is stopped, the paddle-wheels are simply thrown out of gear, and the engine continues to work. The necessary supply of water is thus pumped into the boiler during the whole time that the vessel may be at rest, and when she is required to get under weigh, the wheels are again thrown into gear, and revolve with the paddle-wheel shaft. The fly-wheel, formerly noticed, is useful in regulating the motion of the engine, which otherwise might be apt to suffer damage from the increase and diminution in the resistance offered to the motion of the pistons, by suddenly throwing the paddle-wheels into and out of gear. The water for the supply of the engine is first pumped into a heater, in which its temperature is raised, and is then injected into the boiler.

I saw several vessels on the Ohio which were propelled by one large paddle-wheel placed at the stern of the vessel. but it is doubtful whether this arrangement is advantageous, as the action of the paddle-wheel, when placed in that situation, must be impeded by the floatboards impinging on water which has been disturbed by the passage of the vessel through it.

The Mississippi steamers carry a captain, a clerk, two engineers, and two pilots, one of whom is always at the helm. The firemen and the crew are people of colour, and generally slaves. The passage from New Orleans to Pittsburg, against the current of the river, is generally performed in from fifteen to twenty days, and from Pittsburg to New Orleans in about ten days. The distance is rather more than 2000 miles, and the cabin-passage, including all expenses, is about L.10.

The third class of vessels to which I have alluded, are those which navigate the Lakes and the River St Lawrence. They differ very materially from those I have already described, being more like the steamers of this country, both in their construction and appearance. Steam-boats were first used on the St Lawrence in 1812, and it is probable that they were also introduced on the Lakes about the same time. The Lake steamers are strongly built vessels, furnished with masts and sails, and propelled by powerful engines, some of which act on the high-pressure and some on the low-pressure principle.

The largest steamer on the Lakes in 1837 was the "James Madison." She measures 181 feet in length on the deck, 30 feet in breadth of beam, and 12 feet 6 inches in depth of hold. She carries about 700 tons of goods, and draws about 10 feet of water. This vessel plies between Buffalo on Lake Erie and Chicago on Lake Michigan, a distance of 950 miles. The hulls of the vessels are built in the ports on the shores of the Lakes, and the engines are generally made at Pittsburg. It is somewhat curious to find such vessels engaged in inland navigation; but their dimensions and strength are rendered necessary by the severe storms and formidable waves encountered on the Lakes, to which I have already particularly alluded, in the chapter on Lake Navigation.

Some of the St Lawrence steam-boats, all of which are owned by her Majesty's subjects resident in Canada, are fine powerful vessels. The machinery of most of them is made at Montreal. The "John Bull" is the largest of these vessels, and measures 210 feet in length of deck, 33 feet 6 inches in breadth of beam, and draws 10 feet of water. She is propelled by two condensing engines, having cylinders 60 inches in diameter, and 8 feet in length of stroke. This steamer is principally employed in towing vessels; and of her performance in this way I have already spoken at page 88. She has a small engine of about 3 horses power for pumping water into the boilers while the vessel is at rest.

The vapour contained in the boiler of a steam-engine is liable to have its volume increased or diminished to a dangerous extent by sudden variations of temperature, and the application of an apparatus capable of counteracting the tendency of such changes of temperature to produce rupture, is absolutely indispensable to the safe operation of the boiler. The want of the ordinary precautions necessary for insuring safety, or the inefficient manner in which these are applied, together with the very high pressure at which the vapour is used for propelling the engines of many of the American steam-boats, and the recklessness of the engineers employed on some navigations, have occasioned many disastrous accidents in that country from the explosion of steam-boilers. These, however, as already stated, are now happily, in a great measure, confined to the vessels employed on the Western Waters. The frequent occurrence of these accidents, and the melancholy consequences attending them, induced the Government of the United States in 1832, to institute an inquiry into "the causes of steam-boat explosions, and the best means of preventing them." At that period a list of the explosions which had taken place was made up by Mr Redfield of New York, which I shall give at full length, as the best means of affording an idea of their extent and serious nature.

List of Steam-boat Explosions which have occurred in the United States, by W. C. Redfield.

When exploded.	Names.	Place of Explosion.	Killed.	Wounded.
High Pressure.				
1817	Constitution, .	Mississippi, .	13	0
...	General Robinson,	Do. .	9	0
...	Yankee, . .	Do. .	4	0
...	Heriot, . .	Do. .	1	0
1824	Etna, . . .	New York Bay,	13	0
1828	Grampus, . .	Mississippi, .	unknown	0
...	Barnet, . .	Long Island Sound,	1	0
1830	Helen Macgregor,	Mississippi, .	33	14
...	Caledonia, . .	Do. .	11	11
...	Car of Commerce,	Ohio River, .	28	29
...	Huntress, . .	Mississippi, .	unknown	0
...	Fair Star, . .	Alabama, .	2	0
...	Porpoise, . .	Mississippi, .	unknown	0
			115	54
Low Pressure.				
Previous to 1825	Enterprise, copper boiler, . .	Charleston, S. C.	9	4
...	Paragon, do. .	Hudson River,	1	1
...	Alabama, . .	Mississippi, .	4	0
...	Feliciana, . .	Do. .	2	0
...	Arkansas, . .	Red River, .	4	0
...	Fidelity, copper boiler,	New York Harbour,	2	0
...	Patent, do.	Do. .	5	2
...	Atalanta, do.	Do. .	2	0
...	Bellona, do.	Do. .	2	0
...	Maid of Orleans, do.	Savannah River,	6	0
...	Raritan, unknown,	Raritan, .	1	0
..	Eagle, do.	Chesapeake, .	2	several
...	Bristol, . .	Delaware River,	0	1
...	Powhatan, cop. boiler,	Norfolk, .	2	0
1824	Jersey, do.	Jersey City, .	2	0
1825	Tesch, . .	Mississippi, .	several	0
...	Constitution, .	Hudson River,	3	0
...	Legislator, . .	New York Harbour,	5	2
1826	Hudson, . .	East River, .	0	1
...	Franklin, . .	Hudson River,	1	0
...	Ramapo, in January,	New Orleans, .	5	2
...	Do. in March,	Do. .	1	1
1827	Oliver Ellsworth,	Long Island Sound,	3	0
1830	Carolina, . .	New York Harbour,	1	0
	C. J. Marshal, copper boiler, .	Hudson River,	11	2
	United States, .	Long Island Sound,	9	0
1831	General Jackson,	Hudson River,	12 (supposed)	13
			95	29

N. B.—Of the above low-pressure explosions, ten were copper-boilers, from which were . . . killed 42, wounded 7
8 iron-boilers, ... do. 35, do. 3
9 boilers, metal unknown (probably iron), do. 18, do. 19

The number of copper-boilers in use is now very small compared with those of iron.

Character of Engines not specified.

When exploded.	Names.	Place of Explosion.	Killed.	Wounded.
	Cotton Plant, . .	Mobile, . . .	Unknown.	Unknown.
1816	Washington (high p.)	Ohio River, . .	7	9
1826	Macon, ...	South Carolina, .	4	0
1827	Hornet (low p.), .	Alabama, . . .	2	2
1826	Susquehannah, . .	Susquehannah, .	2	0
1827	Union (high p.), .	Ohio River, . .	4	7
1830	W. Peacock, . .	Buffalo, . . .	15	0
...	Tallyho (high p.), .	Cumberland River,	0	0
...	Kenhawa (low p.),	Ohio River, . .	8	4
...	Atlas, ...	Mississippi, . .	1	0
...	Andrew Jackson, .	Savannah River,	2	0
1831	Tri-color (low p.),	Ohio River, . .	8	8
			46[53?]	21[30?]

Recapitulation.

		Killed.	Wounded.
13	High-pressure accidents, . . .	115	54
27	Low-pressure do.	95	29
12	Character of engines unknown, supposed to be chiefly high pressure, ...	46	21
52	Total,	256	104

"In some of the principal accidents comprised in the foregoing list, the number of killed includes all who did not recover from their wounds. In other cases, the number killed are as given in the newspapers of the day, and some of the wounded should perhaps be added. In some few instances no list has been obtained, and possibly in some no loss of life occurred. The accounts of some of the minor accidents may have been lost sight of. In making an approximate estimate of the whole number of lives which have been lost in the United States by these accidents, I should fix it at 300."

In order to lessen the chances of explosions from the expansive power of the steam, properly constructed boilers are provided with safety-valves, which are loaded with a weight proportioned to the pressure of steam which the boiler is capable of resisting. So long as one of the safety-valves is locked up so as to be inaccessible to the engineers, no danger is to be apprehended from their being overloaded, a practice too frequently resorted to by the ignorant men to whom the management of steam-engines is occasionally entrusted.

The best constructed safety-valves, however, may get deranged from rust or other causes, and by remaining closed after the steam has attained the pressure at which it should be permitted to escape, may fail in performing their duty. A mercurial gauge is generally applied to the boiler, by an examination of which the engineer may at any moment ascertain the expansive power of the steam.

The safety-valves and steam-gauge perform a most important office, and operate chiefly when the engine ceases to work, as, for example, when a steamer stops to land passengers. The volume of vapour which is no longer withdrawn for the supply of the engine, is permitted to escape by the opening of the valves; while the steam-gauge, by indicating any increase of pressure, gives timely warning of danger, and calls the attention of those in charge to such measures as may arrest too rapid accumulation of steam within the boiler. Thus far the safety-valves and steam-gauge have the effect of insuring the safety of the boiler, but unfortunately they have no control over the accidents arising from a deficiency in the supply of water, to which circumstance almost all the explosions which now take place may be traced.

The heat to which the flues and bottom of a steam-boiler are exposed may be very intense, but the metal of which they are formed will preserve a comparatively low degree of temperature, so long as its interior surface is kept in contact with the water. If the level of the water be permitted to sink, however, so as to uncover or lay bare part of the flues or bottom, the action of the fire immediately renders the parts so exposed red hot. When this state of things occurs, a boiler, as we shall presently see, is placed in a most critical situation. Deficiency of water may occur when a vessel is in motion, from derangement of the apparatus for its supply, but it is most apt to arise when a vessel stops for the purpose of taking in goods or landing passengers. On such occasions the working of the engine is stopped, and at the same time the pump for supplying the boiler with water must cease to act. Meanwhile, the fire is kept briskly burning, and if the stoppage is of long duration, the level of the water, from the evaporation which is going on, falls considerably, and occasionally to such an extent that the flues become exposed and are quickly rendered red hot. When the vessel is about to proceed on her voyage the engine is set in motion, and the pump, which has till then remained inactive, injects heated water into the boiler. This water comes in contact with the portions of its surface which have been uncovered and rendered red hot, and is instantaneously converted into vapour. So rapid is this change, resembling in effect the ignition of gunpowder, that the safety-valves, in most instances, are too small to give vent to the immense volume of vapour which is suddenly created, and an explosion of the boiler is the unavoidable consequence.

A proper uninterrupted supply of water is the only safe-guard against the occurrence of such explosions, which, from their nature, are equally apt to occur to low-pressure and high-pressure boilers. Some engines have self-acting pumps for the supply of water, and in others the injection-cock is under the control of the engineer, who by opening or shutting it, regulates the supply. The latter plan is adopted in all locomotive engines, and in most of the American steam-boats. It is of the greatest consequence that the water-pump should be so arranged as to work while the engine is at rest. The steam boats on the eastern part of the United States, are not so constructed; but in the steam-boats on the Mississippi and the St Lawrence, as formerly noticed, I found apparatus for effecting this important object. A gauge is applied to almost every boiler, for indicating the height at which the water stands in its interior, and if this is carefully observed and tried from time to time by the engineer, it forms a great means of preventing accident. Some ingenious applications have been proposed to render the safety of the boiler less dependent on the attention of the workmen. One of these is a valve of larger dimensions than the common safety-valve, which is intended to be acted on by the expansive force of a rod of iron, when heated beyond a certain temperature. The introduction of plates into the sides of the boiler, composed of an easily fusible metal, which would melt before the contained steam had attained a dangerously high temperature, and form large vents for its escape, is another method not unworthy of attention.

The collapse of the large boilers of weak construction, which are sometimes employed for generating low-pressure steam, is another casualty to which steam-vessels are liable. It is occasioned by the fire getting low, and the surface of the boiler becoming cool. This produces condensation of the steam, and the formation of a partial vacuum in the interior of the boiler, the form of which is generally so ill calculated for resisting external pressure, that it yields to the weight of the atmosphere. A spring valve so constructed as to be opened by external pressure alone, is occasionally applied in this country. When a vacuum is formed in the boiler, the valve is opened by the weight of the atmosphere on its exterior surface, and the air rushing in, restores the equilibrium, and insures the safety of the boiler. The exposed situation in which the boilers of all the American steam-boats are placed, renders them very liable to collapse, which has been of very frequent occurrence, and has on some occasions been attended with serious consequences.

Of the several adaptations for reducing the chances of accident which I have mentioned, I found in use in the American steam-boats the single safety-valve, the steam-gauge, and the water-gauge, and in a few vessels the apparatus for continuing the supply of water while the vessel is at rest.

It appears from Mr Redfield's list of accidents, that there have been nearly four explosions every year for the last fourteen years, and an annual loss of twenty-one lives from these accidents. Of the forty cases regarding which definite information had been obtained, twenty-seven were low-pressure engines, and only thirteen high pressure. The average loss of lives by each low-pressure accident, is only three and a half, but the loss by high-pressure accidents averages nine on each occasion. This may be accounted for by the great elasticity of the steam in all the high-pressure engines in America, which in its escape causes proportionally greater mischief.

The following table, containing the dimensions of several of the best steamers plying in America in 1837, was compiled partly from actual measurement of the vessels, and partly from the report of the engineers in charge of them. To Mr Alfred Stillman of New York, I am indebted for much assistance in obtaining the information contained in it.

Dimensions of American Steam-Boats plying in 1837.

Names.	Length of Deck. Ft.	In.	Breadth of Beam. Ft.	In.	Depth of Hold. Ft.	In.	No. of Engines.	Length of Stroke. Ft.	Diam. of Piston. In.	Diameter of Paddle-wheels. Ft.	In.	Breadth of do. Ft.	Dip of do. In.	Draft of Water. Ft.	At what part of Stroke cuts off the Steam.	No. of Double Strokes ⅌ minute.	Remarks.
Dewit Clinton,	235	0	28	0	9	0	1	10	66	22	0	14	28	6	¾ from bottom of cylinder.	...	An old boat, plying between New York and Albany.
Providence,	180	0	27	0	...		1	10	65	...		...	...	9	...	...	An old boat, plying between New York and Providence.
Champlain,	180	0	28	0	9	0	2	10	42	22	0	14½	30	...	⅓	26	Two fine vessels, plying between New York & Albany, having 4 boilers placed on the guards, 2 on each side; they burn 35 or 40 cords of wood per trip.
Erie,	180	0	28	0	9	0	2	10	44	22	0	14½	30	...	⅓	26	
North America,	200	0	26	0	8	0	2	8	44	...		...	...	...	...	...	Plying between New York and Albany.
Independence,	148	0	26	0	...		1	10	44	...		...	...	...	...	...	Do. do.
Albany,	212	0	26	0	9	0	1	10	65	24	4	14	30	...	...	19	
Lexington,	207	0	21	0	11	0	1	11	48	23	0	9	30	...	...	24	Do. New York and Providence.
R. L. Stevens,	175	0	24	0	...		1	10	36	22	0	11	...	...	...	...	
Bunkerhill,	...		24	0	9	0	1	11	41	21½	0	11	24	...	...	26	Do. New York and Hartford.
Highlander,	175	0	24	0	8	0	1	10	41	20	0	9	29	...	⅓	29	Do. New York and Newburgh.
Narragansett,	210	0	26	0	10	7	1	11½	56	25	0	11	26	5	½	24	Do. New York and Providence.
Massachusetts,	200	0	30	0	12	0	2	9	44	21½	0	...	...	...	...	...	Do. do.
Rhode Island,	210	0	26	0	...		1	11	60	24	0	11	30	6½	...	21	Do. do.
Swallow,	224	0	22	0	8½	0	1	10	46	22½	0	...	...	...	...	...	Do. New York and Albany.
Rochester,	209	10	24	0	8½	0	1	10	43	24	0	10	30	4	½	27	Do. do.
Giraffe,	175	0	26	0	7½	0	1	11	44	26	0	9	...	4	...	...	Do. New Orleans and Mobile.
Utica,	180	0	21½	0	...		1	10	39	22	0	10	...	...	...	...	Do. New York and Albany.
Winoosky,	135	0	21	0	...		1	7	33	19	0	7	22	4½	½	22	Plying on Lake Champlain.
New York,	228	0	22	8	...		1	10	50	24½	0	12	30	4	½	22	Plying between New York & Newhaven.

CHAPTER V.

FUEL AND MATERIALS.

Fuel used in Steam-Engines and for domestic purposes—Wood—Bituminous Coal—Anthracite Coal—Pennsylvanian Coal-mines—Boilers for the combustion of Anthracite Coal—Building Materials—Brick—Marble—Marble-quarries of New England and Pennsylvania—Granite—Timber—Mode of conducting the "Timber Trade"—" Booms"—Rafts on the St Lawrence, and on the Rhine—Woods chiefly used in America—Live Oak—White Oak—Cedar—Locust—Pine—" Shingles"—Dimensions of American Forest Trees.

I NEED scarcely mention, that wood is very much used as fuel throughout the greater part of the United States and the British dominions in America, both for domestic purposes and for steam-engines, excepting in the neighbourhood of most of the large towns, where, the surrounding country having been cleared and brought into cultivation, it has now become very scarce, and much too valuable to be made use of in that way In such situations coal has of course been substituted in its place. Still, however, throughout a large part of the territory of the United States, the forest is looked to for the great supply of fuel. The firewood is cut into pieces about four feet long, and twelve inches in girth, and is sold in piles four feet square, and eight feet in length, containing each 128 cubic feet, a measure called by the Americans, a "cord." It varies in price in different parts of the country. In New York, a cord of wood costs about 20s.; in Albany, 14s.; on Lake Champlain, the average price is 9s.; on the St Lawrence, 7s. 3d.; and on Lake Ontario, 5s.; its value gradually decreasing as the country becomes less populous. On the Mississippi and Ohio, the price of wood is from 5s. to 8s. a cord. Many experiments have been made in America to ascertain the relative values of wood and coal as fuel for steam-engines; the result of which is, that about two and three-fourth cords of wood, and one ton of coal, generate, in well-constructed boilers, an equal quantity of steam. Pine timber is considered to be the best fuel: its texture is more open, and its combustion is more perfect than hardwood, the heart or interior of which, being less affected by the heat, is often left unconsumed.

An abundant supply of fresh air, and a capacious fire-place, are the great objects to be attained in boilers intended for the combustion of wood. To insure the first of these desiderata, the boilers of the improved steam-boats, as formerly mentioned, are placed on the guards of the vessel. No ash-pit is placed below the fire-grate; and the ashes and charcoal which come from the fire fall directly into the water, while a copious stream of fresh air, constantly ascending through the fire-bars, affords a large supply of oxygen for the combustion of the fuel. The most advantageous depth of the fire-grate, or the space left between the fire-bars and the bottom of the boiler for the reception of the wood, has been found in practice to be about three feet.

Bituminous coal occurs in large quantities on the western side of the Alleghany Mountains, and has been extensively worked in the neighbourhood of Pittsburg, where it is much used in the manufacture of iron. This coal occurs in other parts of the United States, particularly in New England and in Rhode Island. In the British dominions of Nova Scotia, a vein has also been opened at the Albion coal-mines, which is said to be fifty feet in thickness. The steam-boats on the Ohio, and also on the St Lawrence, occasionally burn bituminous coal; but the fire-places are all too large for coal, having been constructed for the combustion of wood.

Anthracite coal has been more extensively worked, and is much more generally used in the United States for domestic purposes, than bituminous coal. The most extensive anthracite coal-fields occur in the State of Pennsylvania, on the courses of the rivers Schuylkill and Lehigh, the navigation of which has been improved at a great expense, to facilitate the carriage of the coal from the mines to the sea for shipment. It has also been found on the banks of the Merrimac, in New England.

The Schuylkill and Lehigh coal-fields lie between

a mountain called the Blue Ridge and the river Susquehanna, and are situate about 100 miles north-east of Philadelphia, the port from which the coal is shipped. The most extensive workings are at Pottsville, on the Schuylkill, and Mauch Chunk, on the Lehigh. At Pottsville, the strata of coal dip from N.E. to S.W., at an angle of about 45°, and at Mauch Chunk they are nearly horizontal. They are in general worked by level drifts, carried into the face of a long range of rising ground, which is entirely composed of one vast bed of coal. The quantity of coal brought from the Pennsylvanian mines to Delaware Bay during the year 1836, was no less than 696,526 tons.

The anthracite coal of North America has a strong resemblance to that found in some parts of Wales, and also in Ireland. It is exceedingly close-grained, has a bright lustre, and, when broken, the fracture presents a great variety of fine colours, from which circumstance it has received in America the name of "peacock-tail" coal. It requires a very high temperature for its combustion, and in order to obtain this, it is necessary that the fire-places in which it is used should be lined with a good non-conducting substance. It has been several times tried in the boilers commonly used in steam-boats, but in the fire-places of the common construction it was found that the coal was brought too closely into contact with the bottom of the boiler and flues, and the caloric being too suddenly withdrawn from it, the fire burned languidly and was occasionally extinguished. Dr Nott of New York has bestowed much labour and time in constructing a boiler and fire-place suited for anthracite coal. These have been introduced in one or two steam-boats, and particularly in some of the ferry-boats plying in the bay of New York. This kind of coal is also burned in the locomotive engines on the Baltimore and Washington railway; but its application to the purpose of generating steam, cannot yet be said to have assumed a more permanent character than that of an experiment.

The principle on which the anthracite boilers are constructed is sufficiently simple. The combustion of the fuel is carried on in a chamber lined with a non-conducting substance, which is quite detached from the boiler, and the heated air only is allowed to pass through the flues, so that the disadvantages arising from the rapid abstraction of caloric from the fuel, which takes place in fire-places constructed for the combustion of bituminous coal or wood, are in this boiler completely obviated. The coal is also broken into small pieces about the size of a hen's egg, and in this way a great surface is exposed to the atmospheric air, and a thorough combustion of the fuel is produced.

The anthracite coal is much used for domestic purposes in New York, Philadelphia, Baltimore and Washington. It is burned sometimes in stoves, and sometimes in an open fire-place. The heat given out by it, when burned in either way, being very dry, evaporating pans are generally used to produce that degree of moisture in the apartments which is requisite to counteract the disagreeable effects produced by breathing a dry and close atmosphere.

Brick is the building material uniformly used for dwelling-houses in the large towns in the United States, in most of which wooden structures are not now permitted to be erected. The public edifices, however, are generally built of marble, which is found in great abundance in different parts of the country.*

Several marble quarries have been opened in Massachusetts and in Vermont, which produce good materials for ordinary building purposes. The City Hall at New York, and the State House at Albany, have been built of the stone produced by these quarries. This marble has a white ground with blue streaks, but its colour lies in irregular patches; and its effect in a building is not good. The finest marble is found in the neighbourhood of Philadelphia, where several quarries have been opened, and are at present extensively worked. This stone, laid down at Philadelphia, costs from 4s. to 7s. per cubic foot, according to its quality. The Bank of the United States, the Philadelphia Bank, the Mint, the Exchange, and many other public edifices in Philadelphia, are built from these quarries,

* I am indebted to Mr Struthers of Philadelphia for some interesting and valuable information regarding the marbles of the United States.

which produce pure white marble of very good quality. The public buildings in Philadelphia, most of which were designed by Mr Strickland, architect in that city, present by far the finest specimens of architectural design which are to be met with in the United States, and the extreme purity of the marble of which they are built adds greatly to their general effect. The new Girard College at Philadelphia, designed by Mr Walter, architect, is at present in an advanced state of progress, and promises, when completed, to be a magnificent building. The marble of the United States is rather coarse in the grain, and not very suitable for forming the finely wrought capitals of columns; and the materials of those parts of all the pillars of the public buildings in Philadelphia, were therefore brought from Italy.

I visited some of the quarries in the neighbourhood of Philadelphia, in which the beds of marble dipped from north to south at an inclination of 60° with the horizon. In one of them the quarriers were working a bed fourteen feet in thickness, at a depth of one hundred and twenty feet below the surface. The blocks of marble, some of which weighed twelve tons, are raised to the surface of the ground by means of a horse-gin. A thick layer of common limestone rests on the marble; this is blasted off with gunpowder, and burned for making mortar.

Grey coloured granite, of excellent quality, occurs at Quincy in Massachusetts, and Singsing on the

Hudson. The only hydraulic works in which it has been used are the graving-docks at Boston and Norfolk, which have been already noticed; but it has also been used a good deal in New York for door-lintels and stairs, and latterly it has been introduced for public buildings. The Astor Hotel, the Gaol, and some others, are formed of it.

It is much to be regretted that there are no building materials in the neighbourhood of New York. On examining the ground laid open in some of the railway cuttings in the vicinity of the town, I found it to consist of a stratum of gravel from ten to fifteen feet in depth, with boulder-stones of granite, mica-slate, greenstone, and red sandstone; below this, mica-slate occurs, dipping from north to south at an angle of 45°; but it is not fit for building purposes. This formation occurs on the island of Manhattan, on which the town of New York stands, and also on Long Island, which protects its harbour.

The fine timber which the country produces is much employed in all the public works, and, while it serves in some degree to compensate for the want of stone, it also affords great advantages for ship-building and carpentry, which have been brought to high perfection in America. The lumber trade, as it is called in America, that is to say, the trade in wood, is carried on to a greater or less extent on almost all the American rivers; but on the Mississippi and the St Lawrence it affords employment to a vast number of persons. The chief raftsmen, under whose directions the timber expeditions are conducted, are generally persons of very great intelligence, and often of considerable wealth. Sometimes these men, for the purpose of obtaining wood, purchase a piece of land, which they sell after it has been cleared, but more generally they purchase only the timber from the proprietors of the land on which it grows. The chief raftsman, and his detachment of workmen, repair to the forest about the month of November, and are occupied during the whole of the winter months in felling trees, dressing them into logs, and dragging them with teams of oxen on the hardened snow, with which the country is then covered, to the nearest stream. They live during this period in huts formed of logs. Throughout the whole of the newly cleared districts of America, the houses are built of rough logs. The logs are arranged so as to form the four sides of the hut, and their ends are half-checked into each other in such a manner as to allow of their coming into contact nearly throughout their whole length, and the small interstices which remain are filled up with clay. About the month of May, when the ice leaves the rivers, the logs of timber that have been prepared, and hauled down during winter, are launched into the numerous small streams in the neighbourhood of which they have been cut, and floated down to the larger rivers, where their progress is stopped by what is called a "boom." The boom consists of a line of logs, extending across the whole breadth of the river. These are connected by iron links, and attached to stone piers built at suitable distances in the bed of the stream.

The boom is erected for the purpose of stopping the progress of the logs, which must remain within it till all the timber has left the forest. After this, every raftsman searches out his own timber, which he recognises by the mark he puts on it, and, having formed it into a raft, floats it down the river to its destination.

The boom is generally owned by private individuals, who levy a toll on all the wood collected by it. The toll on the Penobscot river is at the rate of three per cent. on the value of the timber; and the income derived from the boom is about L.300 per annum.

The rafts into which the timber is formed, previous to being floated down the large rivers, are strongly put together. They are furnished with masts and sails, and are steered by means of long oars, which project in front as well as behind them. Wooden houses are built on them for the accommodation of the crew and their families. I have counted upwards of thirty persons working the steering oars of a raft on the St Lawrence; from this some idea may be formed of the number of their inhabitants.

The most hazardous part of the lumberer's business is that of bringing the rafts of wood down the large rivers. If not managed with great skill, they are apt to go to pieces in descending the rapids; and it not unfrequently happens, that the whole labour of one, and sometimes two years, is in this way lost in a moment An old raftsman, with whom I had some conversation on board of one of the steamers on the St Lawrence, informed me that each of the rafts brought down that river contains from L.3000 to L.5000 worth of timber, and that he, on one occasion, lost L.2500 by one raft, which grounded in descending a rapid, and broke up. The safest size for a raft, he said, was from 40,000 to 50,000 square feet of surface; and rafts of that size require about five men to manage them. Some rafts are made, however, which have an area of no less than 300,000 square feet. Rafts are brought to Quebec in great numbers from distances varying from one to twelve hundred miles; and it often happens that six months are occupied in making the passage. They are broken up at Quebec, where the timber is cut up for exportation into planks, deals, or battens, at the numerous saw-mills with which the banks of the St Lawrence are studded for many miles, in the neighbourhood of the town. Sometimes the timber is shipped in the form of logs. The timber-rafts of the Rhine are, perhaps, the only ones in Europe that can be compared to those of the American rivers; but none of those which I have seen on the Rhine were nearly so large as the rafts on the St Lawrence, although some of them were navigated by a greater number of hands, a precaution rendered necessary perhaps, by the more intricate navigation of the river.

The woods exported from the St Lawrence are white oak *(Quercus alba)*, the average price of which is 15d. a cubic foot; white pine *(Pinus strobus)*, 4½d.; red pine *(Pinus resinosa)*, 10½d.; elm *(Ulmus Americana)*, 4½d.; and white ash *(Fraxinus acuminata)*, 10d. These, according to the information I received, are the average prices at which the wood sells at Quebec.

The woods used for ship-building in the United States are live-oak *(Quercus virens)*, white oak *(Quercus alba)*, white cedar *(Cupressus thyoides)*, locust *(Robinia pseud-acacia)*, yellow pine *(Pinus variabilis)*, and long-leaved pine *(Pinus palustris*, or *australis* of Michaux*)*.

The live-oak, so called because it is an evergreen, grows only in the Southern States. This valuable wood is too heavy to be applied to a great extent in ship-building, its specific gravity being greater than that of water, and it is generally used along with white oak and cedar for the principal timbers only "The climate becomes mild enough for its growth near Norfolk, in Virginia, though at that place it is less multiplied and less vigorous than in a more southern latitude. From Norfolk it spreads along the coast for a distance of fifteen or eighteen hundred miles, extending beyond the mouth of the Mississippi. The sea air seems essential to its existence, for it is rarely found in the forests upon the mainland, and never more than fifteen or twenty miles from the shore. It is most abundant, most fully developed, and of the best quality, about the bays and creeks, and on the fertile islands which in great numbers lie scattered for several hundred miles along the coast. The live oak is commonly forty or fifty feet in height, and from one to two feet in diameter, but it is sometimes much larger."*

White cedar is considered the most durable wood in use in America. It grows in the Northern States to the height of forty-five or fifty feet, and is sometimes more than ten feet in circumference. The wood is reddish, and somewhat odorous. It is much used in fences, and also for railway sleepers. It does not exist in a natural state in Canada; but the arbor vitæ, which is there called white cedar, is put to all those purposes to which white cedar is applied in the United States. Locust is a hard and durable wood, and is used for treenails. It grows most abundantly in the Southern States; but it is pretty generally diffused throughout the whole country. It sometimes exceeds four feet in diameter, and seventy feet in height. The locust is one of the very few trees that are planted by the Americans. They are often seen forming hedge-rows in the cultivated parts of Pennsylvania. The yellow pine is chiefly confined to the western countries and the range of the Alleghany mountains; and the long-leaved pine is entirely confined to the

* The Sylva Americana. By J. D. Browne, Boston, 1832.

Southern States. These pines are generally employed for the masts and spars of vessels.*

Timber is employed in great quantities in the construction of quays, railways, canal locks, aqueducts, bridges, roofing of houses, and, in short, for every purpose to which it can possibly be applied. The wood used for roofing is formed into pieces called shingles, which measure eighteen inches in length, four inches in breadth, and one-third of an inch in thickness. They are nailed on the rafters of the house, and arranged in the same manner as the slates used in this country. Six inches of each shingle is left exposed to the weather, and as each piece of wood is eighteen inches in length, every part of the roof has three thicknesses of wood, or, in other words, is one inch in thickness. The shingles are generally made of white pine, cedar, or arbor vitæ. They are split with a single blow of the axe, and afterwards smoothed with an instrument resembling a spoke-shave. They cost 8s. per thousand.

The American forests are particularly interesting to the traveller in that country. According to Mr Browne, whose work I have already quoted, there are no less than 140 species of forest-trees indigenous to the United States which exceed 30 feet in height. In

* I would refer such of my readers as desire information regarding the American forest-trees to an excellent paper in the Agricultural Journal for 1835, on the local distribution of trees in the native forests of America, by Mr James Macnab of Edinburgh, who lately made an extensive botanical tour in the United States and Canada.

France there are about thirty, and in Great Britain nearly the same number. One may travel a great way in America without finding a single tree of very large dimensions, but the average size of the trees is far above what is to be met with in this country. The largest tree which I measured was a buttonwood-tree (*Platanus occidentalis*) on the banks of Lake Erie, which I found to be 21 feet in circumference; but I measured very many varying from 15 to 20 feet. M. Michaux mentions, that on a small island in the Ohio, fifteen miles above the mouth of the Muskingum, there was a buttonwood-tree, which, at five feet from the ground, measured 40 feet 4 inches in circumference, giving a diameter of about 13 feet. He mentions another on the right bank of the Ohio, thirty-six miles above Marietta, whose base was swollen in an extraordinary manner; at four feet from the ground it was 47 feet in circumference. This tree ramified at the height of 20 feet from the ground. A buttonwood-tree of equal size is mentioned as existing in Genessee. M. Michaux also measured two trunks of white pine on the river Kennebec, one of which was 154 feet long, and 54 inches in diameter, and the other was 142 feet long, and 44 inches in diameter at three feet from the ground. He also measured one which was 6 feet in diameter, and had reached the greatest height attained by the species its top being about 180 feet from the ground.

CHAPTER VI.

CANALS.

Internal Improvements of North America—Great extent of the Canals and Railways—Introduction of Canals into the United States and Canada—Great length of the American Canals—Small area of their Cross Sections—North Holland Ship Canal—Difference between American and British works—Use of wood very general in America—Wooden Canal-Locks, Aqueducts, &c.—Artificial navigation of the country stopped by ice—Tolls levied, and mode of travelling on the American Canals—Means used in America for forming water-communications—Slackwater navigation on the River Schuylkill, &c.—Construction of Dams, Canals—Locks—Erie Canal—Canal Basin at Albany—Morris Canal—Inclined Planes for Canal lifts, &c.

THE Americans have not rested satisfied with the natural inland navigation afforded by their rivers and lakes, nor made the bounty of Nature a plea for idleness or want of energy; but, on the contrary, they have been zealously engaged in the work of internal improvement; and their country now numbers, among its many wonderful artificial lines of communication, a mountain railway, which, in boldness of design and difficulty of execution, I can compare to no modern works I have ever seen, excepting, perhaps, the passes of the Simplon, and Mont Cenis in Sardinia; but even these remarkable passes, viewed as engineering works, did not strike me as being more wonderful than the Alleghany Railway in the United States.

The objects to which that enterprising people have chiefly directed their exertions for the advancement of their country in the scale of civilization, are the removal of obstructions in navigable rivers; the junction of different tracts of natural navigation; the connection of large towns, and the formation of lines of communication from the Atlantic Ocean to the great lakes, and the valleys of the Mississippi, Missouri, and Ohio. The number and extent of canals and railways which they have executed in effecting these important objects, sufficiently prove that their exertions, during the short time they have been so engaged, have been neither small nor ill directed. The aggregate length of the canals at present in operation in the United States alone, amounts to upwards of two thousand seven hundred miles, and that of the railways already completed to sixteen hundred miles. Nor are the labours of the people at an end, for even now there are no fewer than thirty-three railways in an unfinished state, whose aggregate length, when completed, will amount to upwards of two thousand five hundred miles.

The zeal with which the Americans undertake, and the rapidity with which they carry on every enterprise, which has the enlargement of their trade for its object, cannot fail to strike all who visit the United States as a characteristic of the nation. Forty years ago, that country was almost without a lighthouse, and now no fewer than two hundred are nightly exhibited on its coast; thirty years ago, it had but one steamer and one short canal, and now its rivers and lakes are navigated by between five and six hundred steamers, and its canals are upwards of two thousand seven hundred miles in length; ten years ago, there were but three miles of railway in the country, and now there are no less than sixteen hundred miles in operation. These facts appear much more wonderful when it is considered, that many of these great lines of communication are carried for miles in a trough, as it were, cut through thick and almost impenetrable forests, where it is no uncommon occurrence to travel for a whole day without encountering a village, or even a house, excepting perhaps a few log-huts inhabited by persons connected with the works.

The routes of the principal canals and railroads in North America, which are delineated in the accompanying map, are not wholly confined to the seaward and more thickly peopled States, but extend far into the interior. The stupendous canals which have already been executed enable vessels, suited to the inland navigation of the country, to pass from the Gulf of St Lawrence to the Gulf of Mexico, and also from the city of New York to Quebec on the St Lawrence, or to New Orleans on the Mississippi, without encountering the dangers of the Atlantic Ocean. But, that the reader may be able fully to understand the nature of lines of inland navigation so enormous, I shall give in detail the route from New York to New Orleans, which is constantly made by persons travelling between those places.

	Miles.
From New York to Albany by the River Hudson, the distance is, ...	150
... Albany to Buffalo by the Erie Canal, . .	363
... Buffalo to Cleveland by Lake Erie, . . .	210
... Cleveland to Portsmouth by the Ohio Canal, .	309
... Portsmouth to New Orleans by the Ohio and Mississippi Rivers, ...	1670
Total distance, .	2702

This extraordinary inland journey of no less than 2702 miles, is performed entirely by means of water-communication; 672 miles of the journey are performed on canals, and the remaining 2030 miles of the route is river and lake navigation.

The internal improvements of the United States are placed under the management either of the Legislature of the States in which the works are situate, or of joint-stock companies. The works constructed by the Legislatures of the States are called State-works, and are conducted by commissioners chosen from the different Legislatures, who publish annual reports on the works committed to their charge. The joint-stock companies, on the other hand, are composed of private individuals, who receive a charter from the Government, investing them with power to execute the work, and afterwards to conduct the affairs and transact the business of the company. The public works in the British dominions in North America have been executed partly at the expense, and under the direction of the British Government, and partly by companies of private individuals.

It is believed that canals, which were, until very lately, the only mode of conveyance employed in North America, were in use in Egypt, China, Ceylon, Italy, and Holland, before the Christian era; but the period at which the first artificial water-communication was formed, and the country in which the construction of a canal was first attempted, are equally unknown. The earliest canal constructed in France was the Languedoc, connecting the Bay of Biscay with the Mediterranean Sea, which was completed in the year 1681; and the first formed in Great Britain was that of Sankey Brook in Lancashire completed in 1760. Several short canals were made for improving the river navigation in the United States about the end of the last century; but the first work of any importance in that country was the Santee Canal, in the State of South Carolina, which was opened in the year 1802; and the first in the British dominions in America was the Lachine Canal in Lower Canada, opened in the year 1821. At the end of this chapter, I have given a table of the principal canals in the United States; and their routes, as formerly noticed, are shewn in the map. The table, which is compiled from the American almanacs and the annual reports of the canal commissioners, contains the names of all the canals of any importance now in operation in the country; together with such information, regarding their size and expense, as these documents contain.

The great length of many of the American canals is one remarkable feature in these astonishing works. In this respect they far surpass any thing of the kind hitherto constructed in Europe. The longest canal in Europe is the Languedoc, which has a course of 148 miles; and the most extensive in the United States is the Erie Canal, which is no less than 363 miles in length. But the cross-sectional area of the American canals is by no means so great as that of many in Europe. The North Holland Ship Canal, for example, between the Zuyder Zee, at Amsterdam, and the Helder, which I lately visited, has a larger cross-sectional area than any other European work of the same description. It measures 124 feet 6 inches at the water-line, and affords sufficient breadth to allow large vessels to pass each other with perfect ease. It is 56 feet in breadth at the bottom, and has a depth of water of no less than 21 feet. This remarkable canal, which is nearly fifty miles in length, undoubtedly ranks as one of the greatest works of the kind that has ever been executed. It was constructed for the purpose of facilitating the passage of vessels to and from the port of Amsterdam; and, by means of the sheltered inland passage which it affords, the intricate and dangerous navigation of the Zuyder Zee is avoided. At the time when canals were introduced into America, however, the trade of the country was small, and did not warrant the expenditure of large sums of money in their construction, the chief object being to form a communication with as little loss of time, or outlay of capital, as might be consistent with a due regard for the safety and stability of the work. It is not to be expected, therefore, that the American works, although on an extensive scale, should be constructed in the same spacious style as those of older and more opulent countries. The dimensions of many of the canals in the United States are now found to be inconveniently small for the increased traffic which they have to support; and the great Erie Canal, as well as some others, is at present undergoing extensive alterations, by which its breadth will be increased from 40 to 70 feet, and its depth from 4 to 7 feet. It is doubtful whether the increased depth will, on the whole, prove advantageous, especially for quick transport. According to Mr Russell, the velocity of the wave due to a depth of 4 feet, making allowance for the sloping sides of the canal, is about seven miles an hour; and if the boat is dragged in the top of the wave, the horses must travel at somewhat more than this rate, in order to keep before it. If, on the other hand, the depth of the canal be 7 feet, the velocity of the wave will be about nine miles an hour; a speed which it would be difficult for horses regularly to keep up. The boat would, consequently, travel at a less speed than the wave, which is shewn by Mr Russell, in his Researches in Hydrodynamics, to be very disadvantageous.

English and American engineers are guided by the same principles in designing their works; but the different nature of the materials employed in their construction, and the climates and circumstances of the two countries, naturally produce a considerable dissimilarity in the practice of civil-engineers in England and America. At the first view, one is struck with the temporary and apparently unfinished state of many of the American works, and is very apt, before inquiring into the subject, to impute to want of ability what turns out, on investigation, to be a judicious and ingenious arrangement to suit the circumstances of a new country, of which the climate is severe, —a country where stone is scarce and wood is plentiful, and where manual labour is very expensive. It is vain to look to the American works for the finish that characterises those of France, or the stability for which those of Britain are famed. Undressed slopes of cuttings and embankments, roughly built rubble arches, stone parapet-walls coped with timber, and canal-locks wholly constructed of that material, every where offend the eye accustomed to view European workmanship. But it must not be supposed that this arises from want of knowledge of the principles of engineering, or of skill to do them justice in the execution. The use of wood, for example, which may be considered by many as wholly inapplicable to the construction of canal-locks, where it must not only encounter the tear and wear occasioned by the lockage of vessels, but must be subject to the destructive consequences of alternate immersion in water and exposure to the atmosphere, is yet the result of deliberate judgment. The Americans have, in many cases, been induced to use the material of the country, ill adapted though it be in some respects to the purposes to which it is applied, in order to meet the wants of a rising community, by speedily and perhaps superficially completing a work of importance, which would otherwise be delayed, from a want of the means to execute it in a more substantial manner; and although the works are wanting in finish, and even in solidity, they do not fail for many years to serve the purposes for which they were constructed, as efficiently as works of a more lasting description

When the wooden locks on any of the canals begin to shew symptoms of decay, stone structures are generally substituted, and materials suitable for their erection are with ease and expedition conveyed from the part of the country where they are most abundant, by means of the canal itself to which they are to be applied; and thus the less substantial work ultimately becomes the means of facilitating its own improvement, by affording a more easy, cheap, and speedy transport of those durable and expensive materials, without the use of which, perfection is unattainable.

One of the most important advantages of constructing the locks of canals, in new countries such as America, of wood, unquestionably is, that in proportion as improvement advances and greater dimensions or other changes are required, they can be introduced at little cost, and without the mortification of destroying expensive and substantial works of masonry. Some of the locks on the great Erie canal are formed of stone, but had they all been made of wood, it would in all probability have been converted into a ship-canal long ago.

But the locks are not the only parts of the American canals in which wood is used. Aqueducts over ravines or rivers are generally formed of large wooden troughs resting on stone pillars, and even more temporary expedients have been chosen, the ingenuity of which can hardly fail to please those who view them as the means of carrying on improvements, which, but for such contrivances, might be stopped by the want of funds necessary to complete them.

Mr M'Taggart, the resident engineer for the Rideau canal in Canada, gave a good example of the extraordinary expedients often resorted to, by suggesting a very novel scheme for carrying that work across a thickly wooded ravine situate in a part of the country where materials for forming an embankment, or stone for building the piers of an aqueduct, could not be obtained but at a great expense. The plan consisted of cutting across the large trees in the line of the works, at the level of the bottom of the canal, so as to render them fit for supporting a platform on their trunks, and on this platform the trough containing the water of the canal was intended to rest. I am not aware whether this plan was carried into effect, but it is not more extraordinary than many of the schemes to which the Americans have resorted in constructing their public works; and the great traffic sustained by many of them, notwithstanding the temporary and hurried manner in which they are finished, is truly wonderful. The number of boats navigating the Erie Canal in 1836 was no less than 3167, and the average number of lockages 118 per day; facts which clearly prove the efficiency as well as the utility of the work.

With the exception of some few works in the most southern states of the Union, the artificial navigation of North America, as well as that of the northern rivers and lakes, is completely suspended during a period of from three to five months every year. During that time the water is always withdrawn from the canals and feeders. This precaution is absolutely necessary, as the intense frost with which the country is then visited very soon proves destructive to the locks and aqueducts, by the expansion of the water, which, if permitted to remain in them, is speedily converted into a mass of ice.

The rate of travelling which has been adopted on the American canals, the charges for the conveyance of passengers and goods, and the general laws for regulating canal transport, are fixed by the commissioners who have charge of the different works, and are not exactly the same in every State. The following observations, however, regarding the mode of travelling on the Pennsylvania State canals, are generally applicable to all others in the country.

The tolls paid to the State, by the persons who have boats on these canals, are three halfpence per mile for each boat, and three farthings per mile for each passenger conveyed in them. The passenger-boats vary from twelve to fifteen feet in breadth, and are eighty feet in length; the large-sized boats weigh about twenty tons and cost L.250 each, and when loaded with a full complement of passengers draw twelve inches of water. They are dragged by three horses at once, which run ten-mile stages. The length of the tow-line generally used is about 150 feet, and the rate of travelling is from four to four and a half miles per hour.

The canal travelling in many parts of America is conducted with so little regard to the comfort of passengers as to render it a very objectionable conveyance. The Americans place themselves entirely in the power and at the command of the captains of the canal-boats, who often use little discretion or civility in giving their orders, and strangers who are unaccustomed to such usage, and would willingly rebel against their tyranny, are in such cases compelled to be guided by the majority of voices, and quietly to submit to all that takes place, however disagreeable it may be. About eight o'clock in the evening, every one is turned out of the cabin by the captain and his crew, who are occupied for some time after the cabin is cleared, in suspending two rows of cots or hammocks from the ceiling, arranged in three tiers one above another. At nine the whole company is ordered below, when the captain calls the names of the passengers from the way-bill, and at the same time assigns to each his bed, which must immediately be taken possession of by its rightful owner on pain of his being obliged to occupy a place on the floor, should the number of passengers exceed the number of beds, a circumstance of very common occurrence in that locomotive land. I have spent several successive nights in this way, in a cabin only 40 feet long by 11 feet broad, with no less than forty passengers; while the deafening chorus produced by the croaking of the numberless bull-frogs that frequent the American swamps was so great, as to render it often difficult to make one's-self heard in conversation, and, of course, nearly impossible to sleep. The distribution of the beds appears to be generally regulated by the size of the passengers; those that are heaviest being placed in the berths next the floor. The object of this arrangement is partly to ballast the boat properly, and partly, in the event of a breakdown, to render the consequence less disagreeable and dangerous to the unhappy beings in the lower pens. At five o'clock in the morning, all hands are turned out in the same abrupt and discourteous style, and forced to remain on deck in the cold morning air while the hammocks are removed and breakfast is in preparation. This interval is occupied in the duties of the toilette, which is not the least amusing part of the arrangement. A tin vessel is placed at the stern of the boat, which every one washes and fills for his own use from the water of the canal, with a gigantic spoon formed of the same metal; a towel, a brush, and a comb, intended for the general service, hang at the cabin door, the use of which, however, is fortunately quite optional. The breakfast is served between six and seven o'clock, dinner at eleven, and tea at five. The American canal travelling certainly forms a great contrast to that of Holland and Belgium. The boat in which I was conveyed on the canal between Ghent and Bruges, for example, was commodiously fitted up with separate state rooms, containing one berth in each, and was, in other respects, a most comfortable and agreeable conveyance. But I trust the reader will not form an estimate of American travelling from what has just been said, nor take this single specimen of it as a criterion of the whole. In the eastern and earlier settled districts of the country, no such grievances have to be suffered, and there are many hundreds of persons in that part of the United States who hardly believe in their existence. So long as the traveller keeps on the east of the Alleghany Mountains, all goes on smoothly, but if he attempts to cross their summits, and penetrate into the "far west," he must look for treatment such as I have described. There is indeed as great a difference in this respect between the seaward and interior States of North America, as there is between the counties of Kent and Caithness.

But I return from these petty troubles to the consideration of a subject of more importance, namely, the works which have been employed in forming the inland lines of water communication in America. These are of two kinds, called Slackwater navigation and Canals. The Slackwater navigation is the more simple of these operations, and can generally be executed at less expense. It consists in improving the navigation of a river by the erection of dams or mounds built in the stream, which have the effect of damming up the water, and increasing its depth. If there be not a great fall in the bed of the river, a single dam often produces a stagnation in the run of the water, extending for many miles up the river and forming a spacious navigable canal. The tow-path is formed along the margin of the river, and is elevated above the reach of flood water. The dams are passed by means of locks, such as are used in Canals. This method of forming water communication has been extensively and successfully introduced in America, where limited means and abundance of rivers rendered it peculiarly applicable. One of the most extensive works on this principle in the country was constructed by the Schuylkill Navigation Company, in the State of Pennsylvania, and consisted in damming up the water of the river Schuylkill. It extends from Philadelphia to Reading, and is situate in the heart of a country abounding in coal, from the transport of which the Company derives its chief revenue. It is 108 miles in length, and its construction cost about L.500,000. This line of navigation is formed by thirty-four dams thrown across the stream, with twenty-nine locks, which overcome a fall of 610 feet. It is navigated by boats of from fifty to sixty tons burden. These dams are constructed somewhat on the same principle as that erected on the Schuylkill at Fairmount water-works, near Philadelphia. A detailed description of this dam is given in the chapter which treats of water-works.

One great objection to this mode of forming inland navigation, is the necessity of constructing works of great strength, sufficient to enable them to withstand the floods and ice to which they are exposed, and by which they are very apt to be damaged, or even carried away. Accidents of this kind, however, may be in a great measure guarded against by making a judicious selection of situations for the dams and locks, and placing them in such a manner in the bed of the river, that the current may act on them in the direction least detrimental to their stability, as has been done in the dam at Fairmount water-works just alluded to.

The number of boats which passed through the locks of the Schuylkill navigation in 1836, was 24,470, the tolls on which amounted to L.14,043. The various articles taken up the river during that year, weighed 61,079 tons, and those brought towards the sea 570,094 tons, of which 432,045 tons were anthracite coal, from the State of Pennsylvania.

Slackwater navigation also occurs at intervals on many of the great lines of canal. About 78 miles of the Rideau Canal, in Canada, as formerly noticed, are formed in this way, and in the United States it is met with on the Erie, Oswego, Pennsylvania, Frankston, Lycoming, and Lehigh Canals. The works which have been executed in forming most of the water communications in America, however, are not generally of the Slackwater kind, but resemble the canals in use in Europe, being, in fact, artificial trenches or troughs, with locks to enable vessels to pass from one level to another. The locks are furnished with boom-gates, which are opened and shut by a long lever fixed to the tops of the quoin and mitre posts. The sluices by which the water is admitted into the locks, are placed in the lower part of the gates. They are in general common hinge-sluices, opened by means of a rod extending to the top of the gates, and worked by a crank handle.

The canals of this construction in the United States,

are so very numerous, and resemble each other so much, that I do not consider it necessary to give a detailed description of the various works which have been exe cuted on all of them, but shall content myself with giving a brief sketch of the Erie Canal, which was the first in America on which the conveyance of passengers was attempted, and is the longest canal in the world regarding which we possess accurate information.

The Erie Canal was commenced in 1817, and completed in 1825. The main line leading from Albany, on the Hudson, to Buffalo, on Lake Erie, measures 363 miles in length, and cost about L.1,400,000 Sterling. The Champlain, Oswego, Chemung Cayuga and Crooked Lake Canals, and some others, join the main line, and, including these branch canals, it measures 543 miles in length, and cost upwards of L.2,300,000. This canal is forty feet in breadth at the water line, twenty-eight feet at the bottom, and four feet in depth. Its dimensions have proved too small for the extensive trade which it has to support, and workmen are now employed in raising its banks, so as to increase the depth of water to seven feet, and the extreme breadth of the canal to sixty feet. The country through which it passes is admirably suited for canal navigation, and there are only eighty-four locks on the main line. These locks are each ninety feet in length, and fifteen in breadth, and have an average lift of eight feet two inches. The total rise and fall is 692 feet. The tow-path is elevated four feet above the level of the water, and is ten feet in breadth. The Erie Canal begins at Buffalo, on Lake Erie, and extends for a distance of about ten miles along the banks of Lake Erie and the river Niagara, as far as Tonewanta Creek. By means of the Slackwater navigation formerly described, the channel of the Tonewanta is rendered navigable for the distance of twelve miles, and the canal is then carried through a deep cutting, extending seven and a half miles, to Lockport. Here it descends sixty feet by means of five locks excavated in solid rock, and afterwards proceeds on an uniform level for a distance of sixty-three miles to Genessee River, over which it is carried on an aqueduct having nine arches of fifty feet span each. Eight and a half miles from this point it passes over the Cayuga marsh, on an embankment two miles in length, and in some places seventy feet in height. It then passes through Lakeport and Syracuse, and at this place the "long level" commences, which extends for a distance of no less than sixty-nine and a half miles to Frankfort, without an intervening lock. After leaving Frankfort, the canal crosses the river Mohawk, first by an aqueduct of 748 feet in length, supported on sixteen piers, elevated twenty-five feet above the surface of the river, and afterwards by another aqueduct 1188 feet in length, and at last reaches the town of Albany.

Albany is the capital of the State of New York,

and contains a population of about 30,000. It is situate on the west, or right bank of the Hudson, at the head of the natural navigation of the river ; but some improvements have been made, which enable vessels of small burden to ascend as far as Waterford, thirteen miles above Albany. One of these improvements has been effected by the erection of a dam across the Hudson 1100 feet in length and 9 feet in height, at a cost of upwards of L.18,000. The lock connected with this dam measures 114 feet in length and 30 feet in breadth. Albany, however, may be said to monopolize the trade of the river, and, in addition to the interest it possesses as a place of great commerce, it is important from its position at the outlet of the Erie Canal, and as the seat of a large basin or depôt for the accommodation of the boats navigating it. This basin, which has an area of thirty-two acres, is formed by an enormous mound, placed lengthwise with the stream of the River Hudson, and enclosing a part of its surface. The mound is composed chiefly of earth, and is 4300 feet in length and 80 feet in breadth, and being completely covered with large warehouses, it now forms a part of the town of Albany, with which it is connected by means of numerous drawbridges. The place has, in consequence, very much the same appearance as many of the Dutch towns. The lower extremity of the mound is unconnected with the shore, a large passage being left for the ingress and egress of vessels, but its upper end is separated from the bank of the river by a smaller opening, which is closed, when necessary, to prevent ice from injuring the craft lying in the basin. A stream of water is generally allowed to enter at the upper end, which, flowing through the basin, acts as a scour, and prevents it from silting up. The mound is surrounded by a wooden wharf like those of New York and Boston, at which vessels discharge and load their cargoes. This admirable basin forms a part of the Erie Canal works, and cost about L. 26,000.

According to the Report of the Canal Commissioners, dated March 1837, the number of boats registered in the Comptroller's office, as navigating the Erie Canal and its branches, was,—

In 1834,	.	2,585	
... 1835,	.	2,914	Increase, 329
... 1836,	.	3,167	... 253

The total number of clearances or trips made during the same years was,—

In 1834,	.	64,794
... 1835,	.	69,767
... 1836,	.	67,270

The average number of lockages per day at each lock was,—

In 1834,	.	95½
... 1835,	.	112
... 1836,	.	118

The whole tonnage transported on the canal during the year 1836 was 1,310,807 tons, the value of which amounted to 67,643,343 dollars, or L.13,526,868.

The proportion between the weight of freight conveyed from the Hudson to the interior of the country, and that conveyed from the interior of the country to the Hudson, was in the ratio of one to five. The tolls collected in 1836, for the conveyance of goods and passengers, amounted to L.322,867. The rates of charge, according to which the tolls are collected, are annually changed, to suit the circumstances of the trade, and are not the same throughout the whole line of the canal, which renders it difficult to give a view of them. In 1836, the passage-money from Albany to Buffalo in the packet-boat was L.3, 3s., being at the rate of nearly 2d. per mile ; and in a line-boat, which is an inferior conveyance, L.1, 18s., being at the rate of one penny and two-tenths per mile. The expenditure for keeping the canal and its branches in repair during 1836 was 410,236 dollars, or about L.82,047, which, taking the whole length at 543 miles, gives an average of L.151 per mile. The average cost of repairs for the six preceding years amounted to L.136 per mile.

Before leaving the subject of canals, I must not omit to mention the Morris Canal, in the State of New Jersey, which I visited in company with Mr Douglass, the engineer for that work, to whom I am essentially indebted for the information and attention which I received from him during my stay in America. This canal leads from Jersey on the Hudson to Easton on the Delaware, and connects these two

PLATE VI.

Fig. 1.

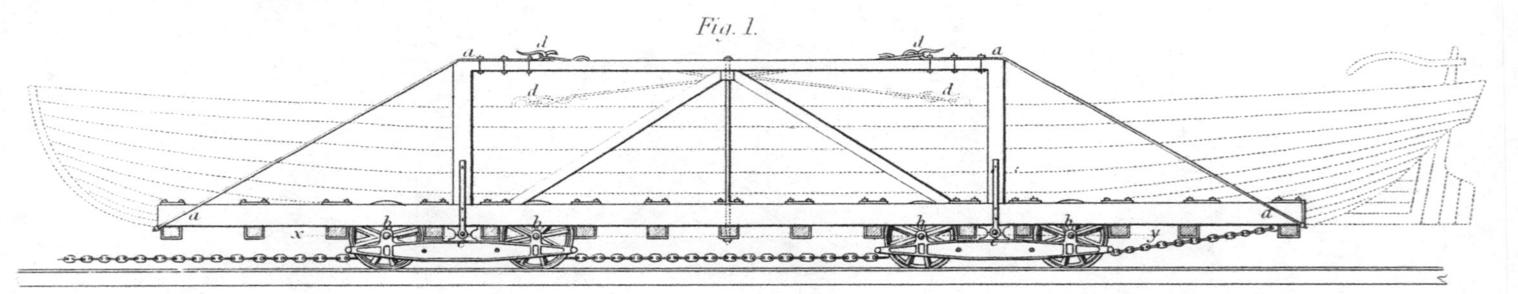

Fig. 2.

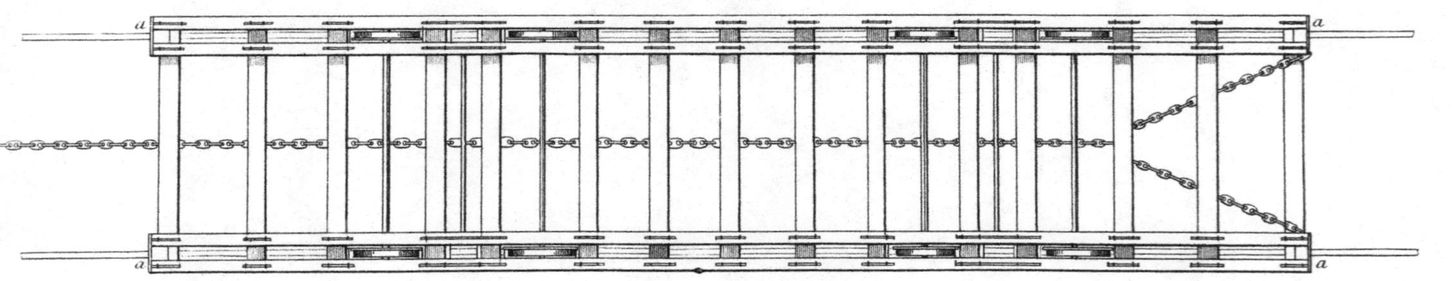

James Andrews, Delt.

Boat car used on the inclined planes at the Morris Canal.

Geo. Aikman, Sculpt.

Published by John Weale, 59, High Holborn, 1838.

rivers. The breadth at the water-line is thirty-two, and at the bottom sixteen feet, and the depth is four feet. It is 101 miles in length, and is said to have cost about L.600,000. It is peculiar as being the only canal in America in which the boats are moved from different levels by means of inclined planes instead of locks; a construction which was first introduced on the Duke of Bridgewater's Canal, in England. The whole rise and fall on the Morris Canal is 1557 feet, of which 223 feet are overcome by locks, and the remaining 1334 feet by means of twenty-three inclined planes, having an average lift of 58 feet each. The boats which navigate this canal are 8½ feet in breadth of beam, from 60 to 80 feet in length, and from twenty-five to thirty tons burden. The greatest weight ever drawn up the planes is about fifty tons. Plate VI. is a drawing of one of the boat-cars used on this canal. Fig. 1 is an elevation, in which the boat is shewn in dotted lines; and fig. 2 is a plan of the car. It consists of a strongly made wooden crib or cradle, marked letter *a*, on which the boat rests, supported on two iron waggons running on four wheels. When the car is wholly supported on the inclined plane, or is resting on a level, the four axles of the waggons, *b b b b*, are all in the same plane, as shewn by the dotted line *x y*; but when one of the waggons rests on the inclined plane, and the other on the level surface, their axles no longer remain in the same plane, and their change of position produces a tendency to rack the cradle, and the boat which it supports; but this has been guarded against in the construction of the boat-cars on the Morris Canal, by introducing two axles, shewn at letters *c c*, on which the whole weight of the crib and boat are supported, and on which the waggons turn as a centre. The cars run on plate-rails laid on the inclined planes, and are raised and lowered by means of machinery driven by water-wheels. I examined several of the planes on this canal near Newark, which appeared to operate remarkably well. The railway, on which the car runs, extends for a short distance from the lower extremity of the plane along the bottom of the canal; when a boat is to be raised, the car is lowered into the water, and the boat being floated over it, is made fast to the part of the framework which projects above the gunwale, as shewn in the drawing at letter *d*. The machinery is then put in motion; and the car bearing the boat, is drawn by a chain to the top of the inclined plane, at which there is a lock for its reception. The lock is furnished with gates at both extremities; after the car has entered it, the gates next the top of the inclined plane are closed, and, those next the canal being opened, the water flows in and floats the boat off the car, when she proceeds on her way. Her place is supplied by a boat travelling in the opposite direction, which enters the lock, and the gates next the canal being closed, and the water run off, she grounds on the car. The gates next the plane are then opened, the car is gently lowered to the bottom when it enters the water, and the boat is again floated. The principal objection urged against the use of inclined planes in canal navigation, for moving boats from different levels, is founded on the injury which the boats are apt to sustain in supporting great weights while resting on the cradle during its passage over the planes. It can hardly be supposed that a slimly built canal boat, measuring from sixty to eighty feet in length, and loaded with a weight of twenty or thirty tons, can be grounded even on a smooth surface, without straining and injuring her timbers; a circumstance which is a decided objection to this mode of construction, and has operated powerfully in preventing its introduction in many situations both in this country and in America. But, notwithstanding this objection, the twenty-three inclined planes on the Morris Canal are in full operation, and act exceedingly well. No pains have been spared to render the machinery connected with them as perfect as possible, and the greatest credit is due to the engineer for the success which has hitherto attended their operation.

The Lachine, the Rideau, the Grenville, and the Welland Canals, and the St Lawrence Canal, at present in progress, are the only artificial water communications in British America; but as I have already noticed these works in the chapters on River and Lake Navigation, it is unnecessary again to allude to them.

TABLE of the Principal Canals and Lines of Slackwater Navigation constructed in the United States up to the year 1836 inclusive. Compiled from the Reports of the Canal Companies, the American Almanac, and other sources.

The Canals marked thus * are State Canals, the others have been executed by Joint Stock Companies.
The Canals marked thus † are shewn on the map.

Names of Canals.	Remarks.	Number of Locks.	Whole Height of Lockage in Feet.	When Opened.	Length in Miles.	Whole length in each State.	Reported Cost.
	Maine.						
Cumberland and Oxford, . .	†From Portland to Sebago Pond, ...	26	. .	1829	20½		£50,000
Songo River, ...	†Slackwater navigation, ...	. .	. .	1829	29½	50	
	New Hampshire.						
Merrimac, ...	†To obviate Falls, on the Merrimac, four in number,	23	121	1812	11	11	28,400
	Massachusetts.						
Middlesex, ...	†Boston Harbour to River Merrimac; breadth at surface 30 feet, at bottom 20 feet, depth 3 feet,	20	136	1808	27		105,600
Blackstone, ...	†Worcestor to Providence, ...	. .	. .	1828	45		120,000
Hampshire and Hampden, .	†Farmington Canal to Northampton, . .	. .	. .	. .	22		
South Hadley, ...	To obviate Falls on the River Connecticut, .	. .	. .	1792	2		
Montague, ...	Do. do. do.	. .	. .	. .	3	99	
	Connecticut.						
Farmington, ...	†Newhaven to the Hampshire and Hampden Canal,	. .	. .	1831	54		600,000
Enfield, ...	To obviate Falls on the River Connecticut, .	. .	. .	. .	5½	59½	
	Carry forward .	. .	. .	. .	. .	219½	

Names of the Canals	Remarks.	Number of Locks.	Whole Height of Lockage in Feet.	Opened.	Length in Miles.	Whole length in each State.	Reported Cost.
	Brought forward,	. .	. .	. .	. .	$219\frac{1}{2}$	
	New York.						
*Erie,	†Albany to Buffalo,	84	. .	1825	363		£1,428,757
*Champlain,	†Albany to Whitehall (with Feeder),	34	. .	1824	76		235,994
*Oswego,	†Syracuse and Oswego (one-half Canal and one-half Slackwater navigation),	14	. .	1828	38		113,087
*Cayuga and Seneca,	†Geneva on Seneca Lake to Montezuma on the Erie Canal,	11	. .	1828	21		47,361
*Chemung,	†Seneca Lake to Chemung River (with Feeder),	53	. .	1833	39		66,338
*Crooked Lake,	Connects Crooked Lake and Seneca Lake,	27	. .	1833	8		
*Chenango,	†Erie Canal and Susquehanna River,	109	. .	. .	97		
Delaware and Hudson,	†Connects the Hudson and Delaware, and extends up the Delaware and Lackawaxen Rivers,	110	1073	1828	109		392,091
Chittenango,	Chittenango to the Erie Canal,	4	. .	. .	$1\frac{1}{2}$		446,364
						$752\frac{1}{2}$	
	New Jersey.						
Morris,	†Jersey to Easton,	. .	1557	1836	101		600,000
Delaware and Raritan,	†Bordentown to New Brunswick (with Feeder),	. .	. .	1834	67		500,000
Salem,	Salem Creek to Delaware,	. .	. .	. .	4		
						172	
	Pennsylvania.						
*Delaware Division of the Pennsylvania Canal,	Bristol to Easton,	. .	164	1830	$59\frac{3}{4}$		247,605
*Central ditto,	†Columbia to Hollidaysburg,	111	585	1830	172		918,829
*Western ditto,	†Johnstown to Pittsburg,	64	470	1830	105		560,000
*Susquehanna ditto,	†Duncan's Island to Northumberland,	. .	86	1831	39		207,851
*North Branch,	†Northumberland to Lackawannock,	. .	111	1830	73		279,682
*West Branch,	Northumberland to Dunnstown,	. .	131	1830	72		316,070
*Beaver,	†Ohio River to Newcastle,	. .	132	. .	25		96,256
*Franklin Line,	†Alleghany River to French Creek (17 miles slackwater navigation),	. .	128	. .	22		88,511
	Carry forward,	. .	. .	. .	$567\frac{3}{4}$	1144	

Names of the Canals.	Remarks.	Number of Locks.	Whole Height of Lockage in Feet.	When Opened.	Length in Miles.	Whole length in each State.	Reported Cost.
	Brought forward,	. .	. .	. .	567¾	1144	
	Pennsylvania—*continued*.						
*French Creek Feeder,	†Bemis Dam to Conneaut Lake,	. .	. .	. .	23		£58,420
Union,	Connecting the rivers Susquehanna and Schuylkill,	91	. .	1827	80		400,000
Schuylkill,	Philadelphia to Reading (slackwater navigation),	129	610	. .	108		500,035
Lehigh,	Delaware River to Stoddartsville (9½ miles slack-water navigation),	53	. .	. .	46½		311,600
Conestoga,	River Susquehanna to Lancaster,	9	. .	. .	16		13,708
Codorus,	. . . .	9	. .	. .	11½		
Conewaga,	To obviate Falls on the Susquehanna,	9	. .	. .	2½		
						855¼	
	Delaware.						
Chesapeake and Delaware,	†Delaware River and Chesapeake Bay (66 feet broad at water line, and 10 feet deep),	. .	. .	1829	14		440,000
						14	
	Maryland.						
Chesapeake and Ohio,	†Finished from Baltimore to Harper's Ferry,	. .	. .	. .	81		
Port Deposit,	To obviate Rapids on the Susquehanna,	. .	. .	. .	10		
Potomac,	To obviate the Falls of the Potomac,	. .	. .	. .	2½		
						93½	
	Virginia.						
Dismal Swamp,	†Chesapeake Bay and Albemarle Sound,	. .	. .	. .	23		175,973
James River,	To obviate Falls on James River,	. .	. .	. .	9½		
						32½	
	North Carolina.						
North-West Canal,	North-West River and Dismal Swamp,	. .	. .	. .	6		
Weldon,	To obviate Falls of the Roanoke,	. .	. .	. .	12		
Lake Drummond Canal,	. . . .	. .	. .	. .	5		
						23	
	Carry forward,	. .	. .	. .	. .	2162¼	

Names of the Canals.	Remarks.	Number of Locks.	Whole Height of Lockage in Feet.	When Opened.	Length in Miles.	Whole length in each State.	Reported Cost.
. . .	Brought forward,	. .	. .	. .	. .	2162¼	
	South Carolina.						
Santee Canal,	†Santee River and Charleston Harbour,	. .	. .	1802	22		£130,133
Wingaw,	†Santee River and Wingaw Bay,	. .	. .	. .	10		
Dreln,	To obviate Fall on Saluda River,	. .	. .	. .	1¼		
Lockhart's,	Shoals on Broad River,	. .	. .	. .	2¾		
Saluda,	Saluda Shoals,	. .	. .	. .	6		
Lorricks,	On Broad River,	. .	. .	. .	1		
Catawha,	To obviate Falls on Catawha River,	. .	. .	. .	11¼		
						54¼	
	Georgia.						
Savannah and Ogeetchee,	†From Savannah to River Ogeetchee,	. .	. .	1829	16		33,000
						16	
	Alabama.						
Huntsville Canal,	†Triana on the Tennessee to Huntsville,	. .	. .	. .	16		
						16	
	Louisiana.						
Carondelet,	†Bayou St John to New Orleans,	. .	. .	1805	6		
Lafourche,	†Navigable only in times of high water,	. .	. .	. .	85		
Lake Veret,	La Fourche Canal to Lake Veret,	. .	. .	. .	8		
						99	
	Kentucky.						
Louisville and Portland,	To obviate Rapids of the Ohio,	4	24	1830	2		
						2	
	Ohio.						
Ohio Canal,	†Lake Erie and Ohio,	152	1205	1832	309		
Miami,	†Cincinnati to Dayton,	32	296	1830	65		149,200
						374	
	Total length,	. .	. .	. .	. .	2723½	

TABLE of the CANALS which are not yet finished, or are proposed to be constructed in the UNITED STATES.

NAMES.	REMARKS.	Length in Miles.	Estimated Cost.
	NEW YORK.		
Genesee and Alleghany,	From the Erie Canal to the Alleghany River,	122½	£378,123
Black River,	From Erie Canal to Black River (begun),	75	
Haerlem,	Across Manhatten Island (begun),	3	110,000
Sodus,	From Seneca River to Lake Ontario,	25	40,000
Scottsville,	From Scottsville to Genesee River,	. .	3,000
Oneida Lake,	From Oneida Lake to Erie Canal (begun),	8½	8,000
Auburn and Owasco,	From Owasco Lake to Auburn,	3	20,000
	PENNSYLVANIA.		
Pittsburg and Erie,	From the Ohio to Lake Erie (begun),	73½	
Chesapeake and Ohio,	From Chesapeake Bay to the River Ohio, (one Tunnel required through the Alleghany Mountains, four miles and eighty yards long, begun),	341¼	1,869,481
	LOUISIANA.		
New Orleans Ship Canal,	... ...	8	70,000
	INDIANA.		
Wabash and Erie,	Maumee River to Lake Erie (begun),	187	
Central,	From Wabash Canal to the Ohio River (begun),	290	
Whitewater,	... ...	76	
Terrehaute and Eel River,	From Wabash Canal to the Central Canal (begun),	40½	125,926
	ILLINOIS.		
Illinois and Michigan,	From Chicago to Illinois River (begun),	95	1,400,000
	ALABAMA.		
Florence,	To overcome the mussel shoals on the Tennessee River (begun),	37	
	OHIO.		
Mahoning and Beaver,	Newcastle to the Ohio Canal (begun),	88	152,874
Sandy Creek,	From the River Ohio to Bolivar (begun).		
	Total length,	1473¼	

CHAPTER VII.

ROADS.

Roads not suitable as a means of communication in America—Condition of the American Roads—" Corduroy Roads"—Road from Pittsburg to Erie—New England Roads—The " National Road" —The " Macadamized Road"—City Roads—Causewaying or Pitching--Brick Pavements--Macadamizing—Tesselated wooden Pavements used in New York and in St Petersburgh.

ROAD-MAKING is a branch of engineering which has been very little cultivated in America, and it was not until the introduction of railways that the Americans entertained the idea of transporting heavy goods by any other means than those afforded by canals and slackwater navigation. Their objection to paved or Macadamized roads such as are used in Europe is founded on the prejudicial effects exerted upon works of that description by the severe and protracted winters by which the country is visited, and also the difficulty and expense of obtaining materials suitable for their construction, and for keeping them in a state of proper repair. Stone fitted for the purposes of road-making is by no means plentiful in America; and as the number of workmen is small in proportion to the quantity of work which is generally going forward in the country, manual labour is very expensive. Under these circumstances, it is evident that roads would have been a very costly means of communication, and as they are not suitable for the transport of heavy goods, the Americans, in commencing their internal improvements, directed their whole attention to the construction of canals, as being much better adapted to supply their wants.

The roads throughout the United States and Canada, are, from these causes, not very numerous, and most of those by which I travelled were in so neglected and wretched a condition, as hardly to deserve the name of highways, being quite unfit for any vehicle but an American stage, and any pilot but an American driver. In many parts of the country, the operation of cutting a track through the forests of a sufficient width to allow vehicles to pass each other, is all that has been done towards the formation of a road. The roots of the felled trees are often not removed, and in marshes, where the ground is wet and soft, the trees themselves are cut in lengths of about ten or twelve feet, and laid close to each other across the road, to prevent vehicles from sinking, forming what is called in America a "Corduroy road," over which the coach advances by a series of leaps and starts, particularly trying to those accustomed to the comforts of European travelling. The following diagram represents the manner in which these roads are formed,

Fig. 1 being a plan, and Fig. 2 a view of the ends of the logs.

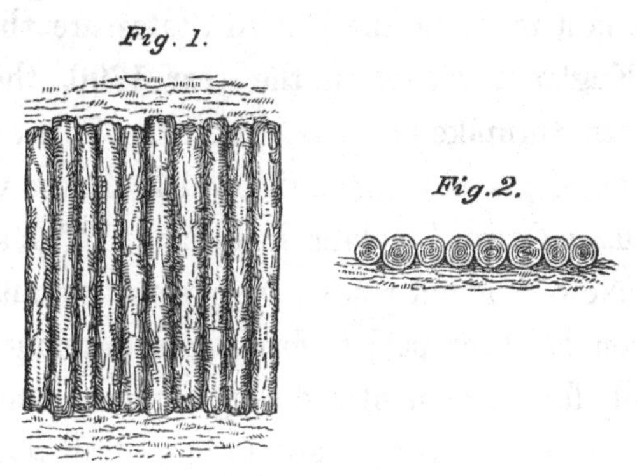

On the road leading from Pittsburg on the Ohio to the town of Erie on the lake of that name, I saw all the varieties of forest road-making in great perfection. Sometimes our way lay for miles through extensive marshes, which we crossed by corduroy-roads, formed in the manner shewn above; at others the coach stuck fast in mud, from which it could be extricated only by the combined efforts of the coachman and passengers; and at one place we travelled for upwards of a quarter of a mile through a forest flooded with water, which stood to the height of several feet on many of the trees, and occasionally covered the naves of the coach-wheels. The distance of the route from Pittsburg to Erie is 128 miles, which was accomplished in forty-six hours, being at the very slow rate of about two miles and three quarters an hour, although the conveyance by which I travelled carried the mail, and stopped only for breakfast, dinner, and tea, but there was considerable delay caused by the coach being once upset and several times "mired."

The best roads in the United States are those of New England, where, in the year 1796, the first American turnpike-act was granted. These roads are made of gravel; a material which, by the way, is much used for road-making in Ireland. The surface of the New England roads is very smooth; but as no attention has been paid to forming or draining them, it is only for a few months during summer that they possess any superiority, or are, in fact, at all tolerable. In Virginia and all the States lying to the south, as well as throughout the whole country to the westward of the Alleghany Mountains, the roads, I believe, are, generally speaking, of the same description as the one already mentioned between Pittsburg and Erie, affording very little comfort or facility to those who have the misfortune to be obliged to travel upon them.

But on the construction of one or two lines of road, the Americans have bestowed a little more attention. The most remarkable of them is that called the "National Road," stretching across the country from Baltimore to the State of Illinois, a distance of no less than 700 miles, an arduous and extensive work, which was constructed at the expense of the government of the United States. The narrow tract of land from which it was necessary to remove the timber and brushwood for the passage of the road, measures eighty feet in breadth; but the breadth of the road itself is only thirty feet. The line of the "National Road" is laid down on the accompanying map. Commencing at Baltimore, it passes through part of the State of Maryland, and entering that of Pennsylvania, crosses the range of the Alleghany Mountains, after which, it passes through the States of Virginia, Ohio and Indiana, to Illinois. It is in contemplation to produce this line of road to the Mississippi at St Louis, where, the river being crossed by a ferry-boat stationed at that place, the road is ultimately to be extended into the State of Missouri, which lies to the west of the Mississippi.

The "Macadamized road," as it is called, leading from Albany to Troy, is another line which has been formed at some cost, and with some degree of care. This road, as its name implies, is constructed with stone broken, according to Macadam's principle. It is six miles in length, and has been formed of a sufficient breadth to allow three carriages to stand abreast on it at once. It belongs to an incorporated company, who are said to have expended about L.20,000 in constructing and upholding it.

Some interesting experiments have lately been set on foot at New York, for the purpose of obtaining a permanent and durable City Road, for streets over which there is a great thoroughfare. The place chosen for the trial was the Broadway, in which the traffic is constant and extensive.

The specimen of road-making first put to the test

was a species of causewaying or pitching; but the materials employed are round water-worn stones, of small size; and their only recommendation for such a work appears to be their great abundance in the neighbourhood of the town. The most of the streets in New York, and indeed in all the American towns, are paved with stones of this description; but, owing to their small size and round form, they easily yield to the pressure of carriages passing over them, and produce the large ruts and holes for which American thoroughfares are famed. To form a smooth and durable pavement, the pitching-stones should have a considerable depth, and their opposite sides ought to be as nearly parallel as possible, or, in other words, the stones should have very little taper. The footpaths in most of the towns are paved with bricks set on edge, and bedded in sand, similar to the "clinkers," or small hard-burned bricks so generally used for road-making in Holland.

The second specimen was formed with broken stones, but the materials, owing chiefly no doubt to the high rate of wages, are not broken sufficiently small to entitle it to the name of a "Macadamized Road." It is, however, a wonderful improvement on the ordinary pitched pavement of the country, and the only objections to its general introduction, as already noticed, are the prejudicial effects produced on it by the very intense frost with which the country is visited, and the expense of keeping it in repair.

The third specimen is rather of an original description. It consists of a species of tesselated pavement, formed of hexagonal billets of pine wood measuring six inches on each side, and twelve inches in depth, arranged as shewn in the following cut, in which Fig. 3

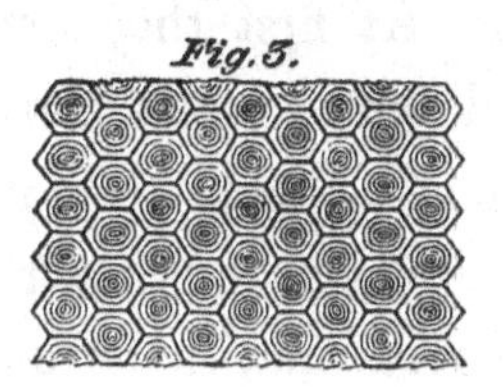
Fig. 3.

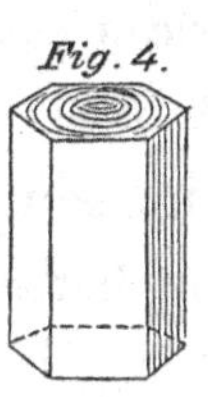
Fig. 4.

is a view of part of the surface of the pavement, and Fig. 4 is one of the billets of wood of which it is composed, shewn on a larger scale. From the manner in which the timber is arranged, the pressure falls on it parallel to the direction in which its fibres lie, so that the tendency to wear is very small. The blocks are coated with pitch or tar, and are set in sand, forming a smooth surface for carriages, which pass easily and noiselessly over it. There can be no doubt of the suitableness of wood for forming a roadway ; and such an improvement is certainly much wanted in all American towns, and in none of them more than in New York. Some, however, have expressed a fear that great difficulty would be experienced in keeping pavements constructed in this manner in a clean state, and that during damp weather a vapour might arise from the timber, which, if it were brought into general use, would prove hurtful to the salubrity of large towns.

In the northern parts of Germany and also in Russia, wooden pavements are a good deal used. My friend Dr D. B. Reid informs me, that at St Petersburgh a wooden causeway has been tried with considerable success. The billets of wood are hexagonal, and are arranged in the manner represented in the diagram of the American pavement. At first they were simply imbedded in the ground, but a great improvement has been introduced by placing them on a flooring of planks laid horizontally, so as to prevent them from sinking unequally. This has not, so far as I know, been done in America.

CHAPTER VIII.

BRIDGES.

Great Extent of many of the American Bridges—Different Constructions adopted in America—Bridges over the Delaware at Trenton, the Schuylkill at Philadelphia, the Susquehanna at Columbia, the Rapids at the Falls of Niagara, &c.—Town's "Patent Lattice Bridge"—Long's "Patent Truss Bridge."

THE vast rivers, lakes and arms of the sea, spanned by many of the American bridges, are on a scale which far surpasses the comparatively insignificant streams of this country, and, but for the facilities afforded for bridge-building by the great abundance of timber, the only communication across most of the American waters must still have been by means of a ferry or a ford. The bridge over the river Susquehanna at Columbia, and that over the Potomac at Washington, for example, are each one mile and a quarter in length; and in the neighbourhood of Boston there are no less than seven bridges, varying from 1500 feet to one mile and a half in length. The bridge over Lake Cayuga is one mile, and those at Kingston on Lake Ontario, and at St John's on Lake Champlain, are each more than one-third of a mile in length.

The American bridges are in general constructed entirely of wood. Although good building materials had been plentiful in every part of the country, the consumption of time and money attending the construction of stone-bridges of so great extent must, if not in all, at least in most cases, have proved too considerable to warrant their erection. Many of those recently built, however, consist of a wooden superstructure resting on stone-piers, and in general exhibit specimens of good carpentry, and not unfrequently of good engineering. In those bridges which are of considerable extent and importance, the roadway, and the timbers by which it is supported, are generally protected by a roof or covering to preserve the wood from decay, in the manner shewn in Plate VIII., in which one-half of the bridge is represented as covered in, and the other half as left exposed, in order to shew the timbers. The roadway is lighted by windows, formed at convenient distances in the covering, as shewn in the drawings. The wooden bridges in Switzerland and Germany are generally covered in the same manner as those in America; and by adopting this plan, the objections to wood as a building material, arising from its tendency to decay by exposure to the atmosphere, are in some degree palliated. The planking or flooring of the American bridges is never covered with any composition, as is generally the case in this country, but is left quite bare.

The simplest method of constructing wooden bridges

Bridge over the Schuylkill, at Philadelphia.

Fig. 1.

Water Line.

Scale of Feet.

10 5 0 10 20 30 40 50 60 70 80 90 100 200

Fig. 2.

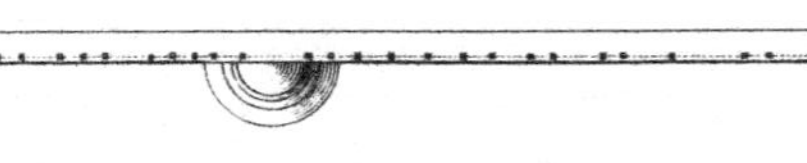

Fig. 3.

Water Line

Stevenson's Sketch of the Civil Engineering of North America.

James Andrews, Delt.

Geo. Aikman, Sculpt.

Published by John Weale, 59, High Holborn, 1838.

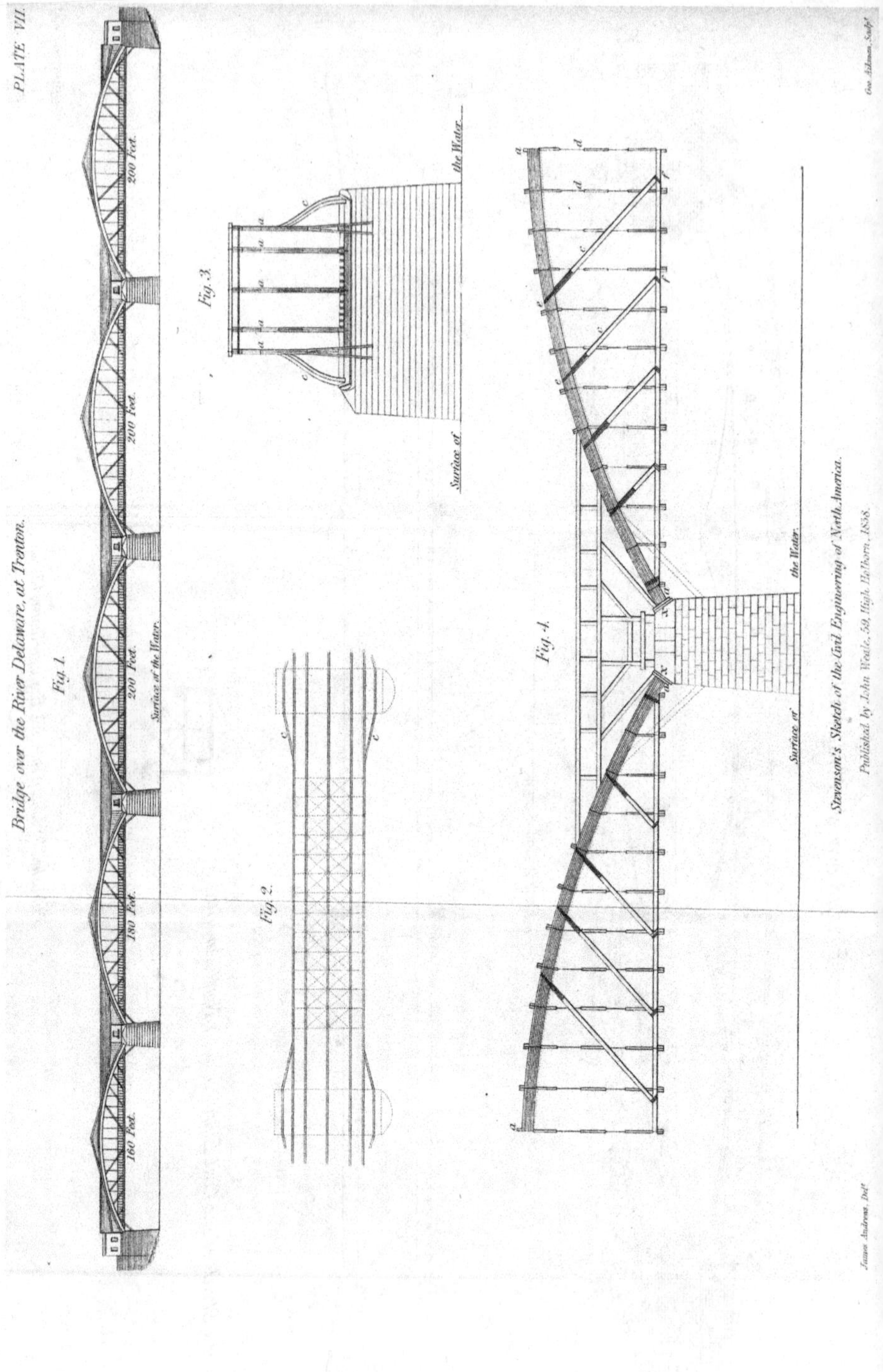
PLATE VII.
Bridge over the River Delaware, at Trenton.
Fig. 1.
160 Feet.
180 Feet.
200 Feet.
200 Feet.
200 Feet.
Surface of the Water.
Fig. 2.
Fig. 3.
Surface of the Water.
Fig. 4.
Surface of the Water.
Stevenson's Sketch of the Civil Engineering of North America.
Published by John Weale, 59, High Holborn, 1838.
James Andrews, Delt
Geo. Aikman, Sculpt

The material originally positioned here is too large for reproduction in this reissue. A PDF can be downloaded from the web address given on page iv of this book, by clicking on 'Resources Available'.

is to form the roadway on horizontal beams, supported on a series of piles driven into the ground, and where the nature of the situation admits of this construction, it is very generally adopted in America. But in spanning rivers, where it is of consequence to preserve a large water way for the passage of ice, or on railways, where it is often necessary that the surface of the rails should have a considerable elevation above the level of the water or ravine over which they are to pass, the use of horizontal beams supported on piles is often wholly impracticable, and in such situations other constructions have been resorted to for forming communications, some of which I shall briefly notice.

Plate VII. is the bridge over the river Delaware at Trenton, about thirty miles from Philadelphia. This bridge consists of five wooden arches, three of 200, one of 180, and one of 160 feet span, supported on four stone piers.* Fig. 1 is an elevation of the bridge, Fig. 2 is a plan of one of the arches, and Fig. 3 is a cross section; Fig. 4 is an enlarged view, shewing one of the piers, and a part of two of the arches. The roadway of each span or opening, is suspended by iron rods, from five wooden arcs, represented by the letter *a* in Figs. 3 and 4, on the same principle as the iron bridge over the river Aire at Leeds in Yorkshire. The wooden arcs in the three largest openings are 200 feet in span, and have a versed sine of 27 feet. The arcs and sus-

* The dimensions of this bridge are not from measurements made by myself.

pending rods divide the roadway into four compartments, as shewn in Fig. 3, forming two carriage-ways in the middle of the bridge, each of which is nine feet ten inches in the clear, and a footpath at each side four feet ten inches in the clear. The entire breadth of the bridge, measured over the outer suspending arcs, is thirty-three feet eight inches. The whole is covered with a roof, in the manner shewn in the drawing.

The suspending arcs, marked letter *a* Fig. 4, butt against strong oak planks, as shewn at letter *x*, which extend throughout the whole breadth of the stone-piers. They are supported at each pier by struts marked letter *c* in Figs. 2 and 3, and are connected at the top by a series of diagonal beams, represented by the dotted lines in Fig. 2. These extend only about half-way down the arcs on each side of the crown, so that they do not interfere with the height of the roadway. The suspending arcs are composed of eight thicknesses of pine plank, and measure two feet eight inches in depth, and one foot one inch in breadth. The planks of which they are made measure one foot one inch in breadth, four inches in thickness, and from thirty to fifty feet in length, and are arranged so as to break joint. The wooden braces, marked letter *c*, Fig. 4, are for the purpose of stiffening the roadway. They are fixed at the points, *e*, to the suspending arcs, and at *f* to the longitudinal bearing beams of the roadway by straps of iron. The suspending rods, *d*, are formed of malleable iron, and occur at every sixteen feet in the two exterior arcs, and at every eight feet in the three inner ones, which support the carriage-way.

The bridge over the Susquehanna at Columbia is constructed somewhat on the same principles as the one at Trenton which I have just described. The wooden suspending arcs, however, do not spring from the level of the roadway, but from a point about eight feet below it. In each span of the bridge, therefore, that part of the roadway which is next the springings is supported upon the arcs; and the centre part of it is suspended from them by a framing of wood. This bridge, which was begun in 1832, and completed in 1834, is perhaps the most extensive arched bridge in the world. It is certainly a magnificent work, and its architectural effect is particularly striking. It consists of no less than twenty-nine arches of 200 feet span, supported on two abutments, and twenty-eight piers of masonry, which are founded on rock, at an average depth of six feet below the surface of the water. The water-way of the bridge is 5800 feet; and its whole length, including piers and abutments, is about one mile and a quarter. The bridge is supported by three wooden arcs, forming a double roadway, which is adapted for the passage both of road and railway carriages. There are also two footpaths; which make the whole breadth of the bridge thirty feet. The arcs are formed in two pieces, each measuring seven inches broad by fourteen inches in depth. These are placed nine inches asunder; and the beams composing the wooden framing, by which the roadway is suspended, are placed between them, and fixed by iron bolts passing through the whole.

Plate VIII. is the "Market Street Bridge," over the Schuylkill at Philadelphia. Fig. 1 is an elevation, fig. 2 a plan, and fig. 3 a cross section. It consists of three arches. The span of the centre arch is 194 feet ten inches, and the versed sine is twelve feet. The other two arches are 150 feet in span, and have versed sines of ten feet. The breadth of the roadway is 35 feet. The piers were built with cofferdams, one of them at the depth of 41, and the other at the depth of 21 feet below the surface of the river at high water. The work was commenced in 1801, and completed in 1805; and the expense, which amounted to L.60,000, was defrayed by a company of private individuals. There is another bridge over the Schuylkill at Philadelphia, consisting of a single arch of no less than 320 feet span, having a versed sine of about 38 feet. This bridge has a breadth of roadway of about 30 feet. It has been erected for several years, and is still in good repair and constant use. I regret, however, that I was unable to procure drawings of the wooden ribs or frames of which it is composed, sufficiently detailed and accurate to enable me to lay them before the public.

The bridge across the rapids of the river Niagara is placed only two or three hundred yards from the edge of the great falls. It extends from the American bank of the river to Goat Island, which separates what is called the "American" from the "British fall." The superstructure of the bridge is formed of timber. It is 396 feet in length, and is supported on six piers, formed partly of stone and partly of wood. When I visited the Falls of Niagara in the month of May, the ice carried down from Lake Erie by the rapids of the river, was rushing past the piers of this bridge with a degree of violence that was quite terrific, and seemed every moment to threaten their destruction.

The following very interesting account of this work is given by Captain Hall.*

"The erection of such a bridge at such a place is a wonderful effort of boldness and skill, and does the projector and architect, Judge Porter, the highest honour as an engineer. This is the second bridge of the kind; but the first being built in the still water at the top of the rapids, the enormous sheets of ice, drifted from Lake Erie, soon demolished the work, and carried it over the falls. Judge Porter, however, having observed that the ice in passing along the rapids was speedily broken into small pieces, fixed his second bridge much lower down, at a situation never reached by the large masses of ice.

"The essential difficulty was to establish a foundation for his piers on the bed of a river covered with

* Forty Etchings, from sketches made in North America, with the Camera Lucida, by Captain Basil Hall. Edinburgh, 1830.

huge blocks of stone, and over which a torrent was dashing at the rate of six or seven miles an hour. He first placed two long beams, extending from the shore horizontally forty or fifty feet over the rapids, at the height of six or eight feet, and counter-balanced by a load at the inner ends. These were about two yards asunder; but light planks being laid across, men were enabled to walk along them in safety. Their extremities were next supported by upright bars passed through holes in the ends, and resting on the ground. A strong open frame-work of timber, not unlike a wild beast's cage, but open at top and bottom, was then placed in the water immediately under the ends of the beams. This being loaded with stones, was gradually sunk till some one part of it—no matter which—touched the rocks lying on the bottom. As soon as it was ascertained that this had taken place, the sinking operation was arrested, and a series of strong planks, three inches in thickness, were placed, one after the other, in the river, in an upright position, and touching the inner sides of the frame-work. These planks, or upright posts, were now thrust downwards till they obtained a firm lodgement among the stones at the bottom of the river; and, being then securely bolted to the upper part of the frame-work, might be considered parts of it. As each plank reached to the ground, it acted as a leg, and gave the whole considerable stability, while the water flowed freely through openings about a foot wide, left between the planks.

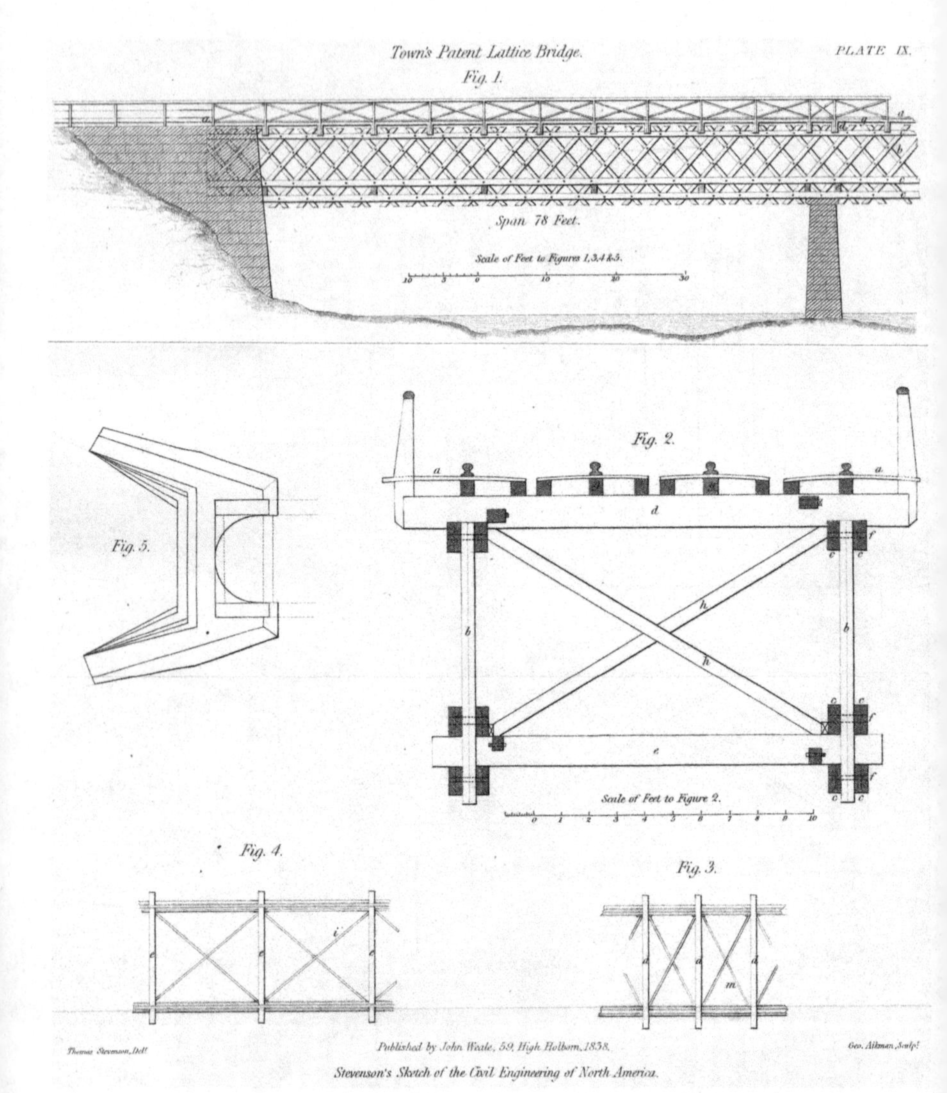

The material originally positioned here is too large for reproduction in this reissue. A PDF can be downloaded from the web address given on page iv of this book, by clicking on 'Resources Available'.

"This great frame or box, being then filled with large stones tumbled in from above, served the purpose of a nucleus to a larger pier built round it, of much stronger timbers firmly bolted together, and so arranged as to form an outer case, distant from the first pier about three feet on all its four sides. The intermediate space between the two frames was then filled up by large masses of rock. This constituted the first pier.

"A second pier was easily built in the same way, by projecting beams from the first one, as had been previously done from the shore; and so on, step by step, till the bridge reached Goat Island. Such is the solidity of these structures, that none of them has ever moved since it was first erected, several years before we saw it."

Plate IX. is a drawing of "Town's Patent Lattice Bridge," which is much employed on the American railways. This construction is sometimes used for bridges of so large a span as 150 feet, and it exerts no lateral thrust tending to overturn the piers on which it rests. A small quantity of materials of very small scantling arranged in the manner shewn in the plate, possesses a great degree of strength and rigidity.

For this drawing I am indebted to Mr Robinson of Philadelphia, who is constructing many large bridges on this principle on the Philadelphia and Reading railway, several of which I examined both in their finished and unfinished state.

Fig. 1 is an elevation, and Fig. 2 a cross section on an enlarged scale of the frame-work of the bridge. The surface of the railway is indicated by letter *a* in both figures. The lattice framing or ribs of which the bridge is formed are composed entirely of pine planks, marked *b*, measuring twelve inches in breadth, and three inches in thickness. The planks are arranged at right angles to each other, so as to form a fabric resembling lattice work, as shewn in the drawing; and from this circumstance the bridge derives its name. They are fixed at the points of their intersection by oak tree-nails, one inch and a half in diameter, passing through them. The horizontal runners, marked *c*, are formed of planks of the same scantling, and extend throughout the whole length of the bridge. They are also fixed at the points where they intersect the planks *b*, by oak treenails passing through the whole, as shewn by the dotted lines at letter *f*, in Fig. 2. The depth of the lattice work is proportioned to the span of the bridge. The span shewn in the drawing is seventy-eight feet, and the depth of the ribs is nine feet six inches. In a bridge of larger span, the planks *b* would be made of greater length, and another square or diamond added to the lattice-work.

There were only two ribs or frames of lattice-work in all of the bridges constructed on this principle which I examined. One of these was placed under each side of the roadway, as shewn in the cross section Fig. 2, by the letters *b b*. The ribs are connected together at the bottom by cross beams marked *e*, at every twelve feet. At the top they are connected in a similar manner by beams marked *d*, at every six feet. On these, the longitudinal beams *g* are supported, to which the planking of the roadway is spiked. To prevent the ribs from twisting or warping, they are braced at every twelve feet by diagonal beams arranged in vertical planes, as shewn at letter *h* in fig. 2. Fig. 3 is a plan of the wood-work directly under the roadway. In this figure the beams *d*, are those on which the planking of the roadway is spiked, and the diagonal braces *m* arranged in horizontal planes are introduced to render the structure rigid. For the same reason the braces *i* are introduced, as represented in Fig. 4, which is a plan of the wood-work connecting the lower part of the lattice frames. The diagonal braces are all fixed in the same manner.— One of the extremities rests in a seat cut for it in the beam against which it butts, and wedges of hardwood are inserted at the other end, by which the brace can be nicely adjusted, and afterwards tightened up, should the vibration of passing trains, or the effects of the atmosphere, cause any yielding of the timber to take place.

The lattice-frames have a rest of about five feet, in checks formed in the stone abutments for their reception, as shewn in dotted lines in the elevation Fig. 1 and in Fig. 5, which is a plan of one of the abutments. If the bridge is of greater extent than can be included in one span, it is simply rested on a thin pier, in the manner shewn in the elevation, without any other support. A covering of light boarding, extending from the level of the roadway to the bottom of the ribs, is spiked on the outside of the lattice-work to preserve the timber.

The largest lattice-bridge which I met with, was constructed by Mr Robinson on the Philadelphia and Reading Railroad. It measures 1100 feet in length. The lattice-frames of which it is formed extend throughout the whole distance between the two abutments without a break, and are supported on ten stone-piers, in the manner shewn in the plate. On the New York and Haerlem Railway, there is a lattice-bridge 736 feet in length, supported in the same manner on four stone-piers.

Plate X. is a drawing of "Long's patent frame bridge," which is also much employed on the different lines of railway in the United States.*

Fig. 1. is an elevation; Fig. 2. a plan; and Fig. 3. a cross section of this bridge, which contains a small quantity of materials, and exerts no lateral thrust. Bridges constructed on this principle, having spans of from one hundred to one hundred and fifty feet, are very commonly met with. That shewn in the drawing is 110 feet in span, and the depth of the truss-frame is 15 feet. The level of the railway is indicated by letter *a* in the Plate; letter *b* represents the

* A Description of Long's Bridge. Concord, 1836.

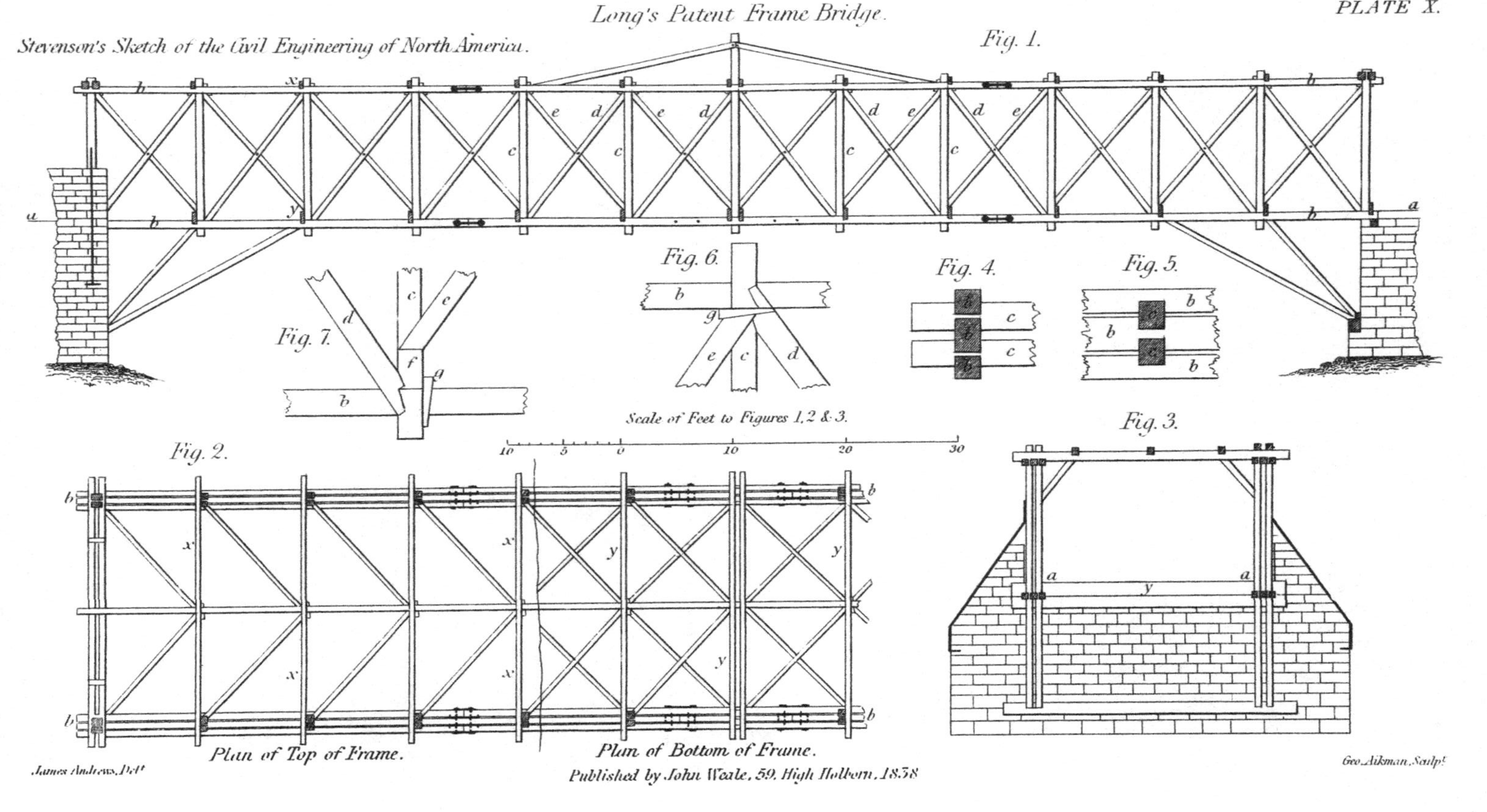

James Andrews, Delt

Geo. Aikman, Sculpt

Published by John Weale, 59, High Holborn, 1838

"string-pieces," as they are called in America; *c* the "posts;" *d* the "main-braces;" and *e* the "counter-braces."

The string-pieces are formed of three beams, in the manner shewn in the plan and cross section. The posts and main-braces are in two pieces, and the counter-beams are formed of a single beam. Figs. 4, 5, 6, and 7, illustrate the manner in which the joining is formed, at the points where the posts and braces are attached to the string-pieces. This joining is effected without the use of bolts or spikes, a construction which admits of the bridge being very easily repaired, when decay of the materials or other causes render it necessary. Figs. 4. and 5. are enlarged diagrams, shewing the manner in which the posts are fixed to the strings. In Fig. 4. the strings are shewn in section at letter *b*, and the posts passing between them at *c*. In Fig. 5. the posts are shewn in section at *c*, and the strings at *b*. Fig. 6. shews the manner of fixing the main and counter braces to the upper string-piece. In this diagram *b* is the string, *c* the post, *d* the main-brace, *e* the counter-brace, and *g* is a wedge of hardwood, by which the whole woodwork is tightened up. Fig. 7 shews the manner of fixing employed at the lower string. In this diagram *b* is the string, *c* the post, *d* the mainbrace, *e* the counter-brace, *g* a wedge of hard-wood, and *f* a block on which the counter-brace rests. The frames are connected at the top by cross beams, *x*, and at the bottom by the beams marked letter *y*, which support the planking of the roadway.

I met with Long's Bridge in many parts of the country, but the best specimens I saw were those erected on some of the railways in the neighbourhood of Boston, under the direction of Mr Fessenden the engineer.

The timbers of which Town's and Long's bridges are composed, are fitted together on the ground previous to their erection on the piers. They are again taken asunder, and each beam is put up separately in the place which it is to occupy, by means of a scaffolding or centering of timber.

CHAPTER IX.

RAILWAYS.

European Railways—Introduction of Railways into the United States—The European construction of Railways unsuitable for America—Attempts of the American Engineers to construct a Railway not likely to be affected by frost—Constructions of the Boston and Lowell, New York and Paterson, Saratoga and Schenectady, Newcastle and Frenchtown, Philadelphia and Columbia, Boston and Providence, Philadelphia and Norristown, New York and Haerlem, Buffalo and Niagara, Camden and Amboy, Brooklyn and Jamaica, and the Charleston and Augusta, Railroads—Rails, Chairs, Blocks, and Sleepers, used in the United States—Original Cost of American Railways—Expense of upholding them—Power employed on the American Railways—Horse-power—Locomotive Engines—Locomotive Engine Works in the United States—Construction of the Engines—Guard used in America—Fuel—Engine for burning Anthracite Coal—Stationary Engines—Description of the Stationary Engines, Inclined Planes, and other works on the Alleghany Railway—Railway from Lake Champlain to the St Lawrence in Canada.

WITHIN a very few years, a wonderful change has been effected in land communication throughout Great Britain and America, where railways have been more extensively and successfully introduced than in any other parts of the world. As early as the sixteenth century, wooden tram-roads were used in the neighbourhood of many of the collieries of Great Britain.

In the year 1767, cast-iron rails were introduced at Colebrookdale, in Shropshire. In 1811, malleable-iron rails were for the first time used in Cumberland, and the locomotive engine, on an improved construction, was successfully introduced on the Liverpool and Manchester line in 1830. Little progress has hitherto been made in the formation of railways on the Continent of Europe. A small one has been in existence for some time in the neighbourhood of Lyons, but the only railroad, constructed in France, for the conveyance of passengers by locomotive power, is that from Paris to St Germains, which was opened only in 1837. In Bohemia, the Chevalier Gerstner, about eight years ago, constructed a railway of eighty miles in length, leading from the river Muldau to the Danube. In Belgium, the railway from Antwerp to Ghent has been in use for some time; and some lines are at present being constructed in Holland and Russia. But my present purpose is to describe the state of this wonderful improvement in communication, in the United States.

The Quincy Railroad in Massachusetts was the first constructed in America. It was intended for the conveyance of stone from the Quincy granite quarries to a shipping port on the river Neponsett, a distance of about four miles. At the end of this chapter I have given a tabular list of the principal railroads which are already finished, and also of those that have been begun in the United States, which shew the rapid increase of these works since 1827, the date at which the

Quincy Railroad was completed. From these tables it appears that, in 1837, there were no fewer than fifty-seven railways completed and in full operation, whose aggregate length amounts to upwards of 1600 miles; and also that thirty-three railways were then in progress, which, when completed, will amount to about 2800 miles. In addition to this, upwards of one hundred and fifty railway companies have been incorporated; and the works of many of them will, in all probability, be very soon commenced.

The early American railroads consisted of iron rails and chairs resting on stone blocks, and were constructed on the same principles as those in this country. But the American engineers soon discovered that this construction of road, although it had been to a certain extent successfully applied in England, was not at all capable of withstanding the rigours of an American winter. The intense frost, with which the northern part of the country is visited, was found to split the stone blocks and to affect the ground in which they were embedded, to such a degree, that their positions were materially altered, and the rails were in many cases so much twisted and deranged as to be quite unfit for the passage of carriages. The consequence was, that most of the railroads constructed in the United States after the English system, had actually to be relaid at the close of every winter, and during the continuance of the frost could only be travelled on at a decreased speed. The Americans have put numerous plans to the test of actual experiment, in their endeavours to form a structure for supporting the rails, adapted to the climate and circumstances of the country. There are hardly two railways in the United States which are made exactly in the same way, and few of them are constructed throughout their whole extent on the same principles ; but although great improvements have undoubtedly been effected, it is doubtful whether a structure perfectly proof against the detrimental effects of frost has yet been produced. An enumeration of the various schemes which have been proposed for the construction of railways in America, would not be very useful, even if it were possible. I shall, therefore, only mention those constructions which came under my own observation, some of which are found to be very suitable.

The Boston and Lowell Railway in Massachusetts is twenty-six miles in length, and is laid with a double line of rails. The breadth between the rails, which is four feet eight and a half inches, is the same in all the American railroads, and the breadth between the tracks is six feet.

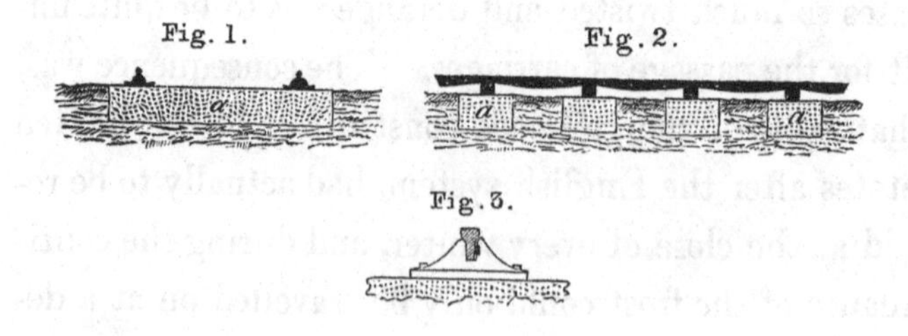

Fig. 1 is a transverse section, and Fig. 2 a side view of one of the tracks, in which *a* are granite blocks six feet in length, and about eighteen inches square. These are placed transversely, at distances of three feet apart from centre to centre, each block giving support to both of the rails. This construction, as formerly noticed by me in some communications made to the Society of Arts for Scotland,* was first introduced in the Dublin and Kingstown Railway, in Ireland, but was found to produce so rigid a road, that great difficulty was experienced in securing the fixtures of the chairs. From the difficulty, also, of procuring a solid bed for stones of so great dimensions, most of them, after being subjected for a short time to the traffic of the railway, were found to be split. The blocks on the Boston and Lowell Railway were affected in the same manner, and are besides found to be very troublesome during frost.

Fig. 3 is an enlarged view of the rail and chair used on this line. The rails are of the kind called fish-bellied. They weigh 40 lb. per lineal yard, and rest in cast-iron chairs, weighing 16 lb. each. The form of the rails and chairs resembles that at first used on the Liverpool and Manchester Railway.

Figs. 4 and 5 represent another construction which has been tried on this line. In these views *a* are longitudinal trenches, two feet six inches square, and

* Transactions of the Society of Arts for Scotland, Edinburgh New Philosophical Journal for April 1835 and April 1836.

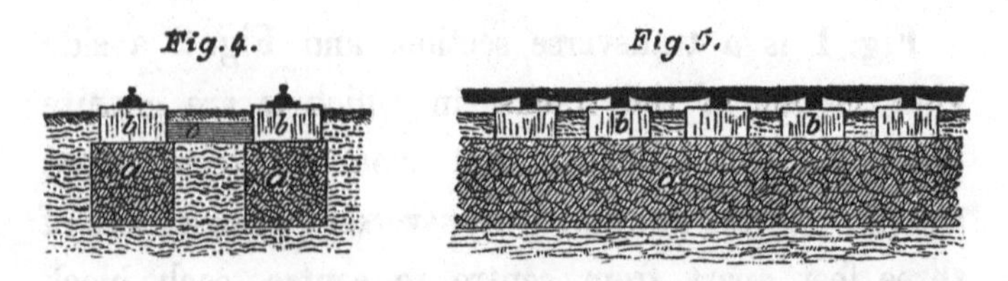

four feet eight and a half inches apart from centre to centre, formed in the ground, and filled with broken stone, hard punned down with a wooden beater, as a foundation for the stone blocks *b* on which the rails rest. These blocks measure two feet square, and a foot in thickness, and *c* is a transverse sleeper of wood, two feet eight inches and a half in length, one foot in breadth, and eight inches in thickness, which is placed between the blocks to prevent them from moving.

The plan of resting the railway on a foundation of broken stone, shewn in the last and some of the following figures, was adopted in the expectation that it might be sunk to a sufficient depth below the surface of the ground, to prevent the frost from affecting it; but it has failed to produce the desired effect, as subsequent experience has shewn that many of those railways whose construction was more superficial have resisted the effects of frost much better.

The New York and Paterson Railway is sixteen and a half miles in length, and extends along a marshy tract of ground. Its construction is shewn in Figs. 6 and 7. The foundation of the road consists of a line of pits under each rail, eighteen inches square,

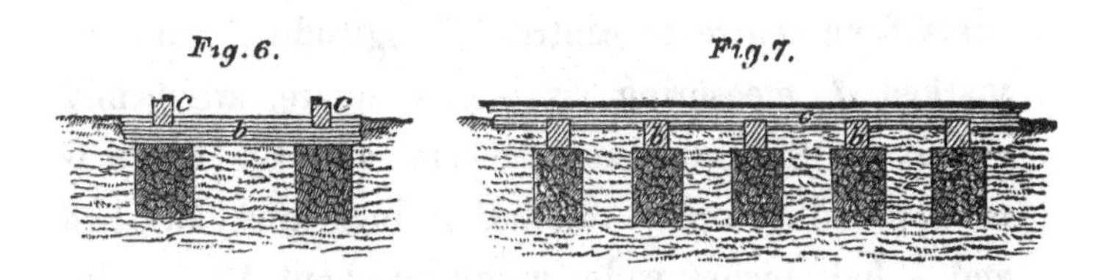

and three feet in depth. They are placed three feet apart from centre to centre, and filled with broken stones. On this foundation transverse wooden sleepers, *b*, measuring eight inches square, and seven feet in length, are firmly bedded, on which rest the longitudinal sleepers marked *c*, measuring eight inches by six. To these, plate-rails of malleable iron, two and a half inches wide, and half an inch thick, weighing about 13 lb per lineal yard, are fixed by iron spikes.

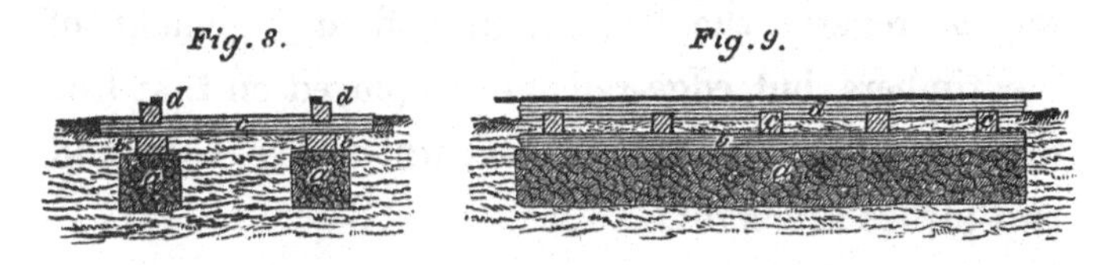

Figs. 8 and 9 are a cross section and side view of the Saratoga and Schenectady Railway. The parallel trenches marked *a*, are eighteen inches square, and four feet eight and a half inches apart from centre to centre. They extend throughout the whole line of the railway, and are firmly punned full of broken stones. Longitudinal sleepers of wood, marked *b*, measuring eight by five inches, are placed on these trenches, which support the transverse wooden sleepers, marked *c*, measuring six inches square, and placed three feet apart from centre to centre. Longitudinal runners, marked *d*, measuring six inches square, are firmly spiked to the transverse sleepers, and the whole is surmounted by a plate-rail half an inch thick, and two and a half inches wide, weighing about 13 lb. per lineal yard.

The Newcastle and Frenchtown Railway, which is sixteen miles in length, and forms part of the route from Philadelphia to Baltimore, is constructed in the same way as that between Schenectady and Saratoga, excepting that the plate-rail is two and a half inches broad, and five-eighths of an inch thick, and weighs nearly 16 lb. per lineal yard. The Baltimore and Washington Railway is also constructed in the same way as regards the foundation and arrangement of the timbers, but edge-rails are employed on that line three and a half inches in breadth at the base, and two inches in height.

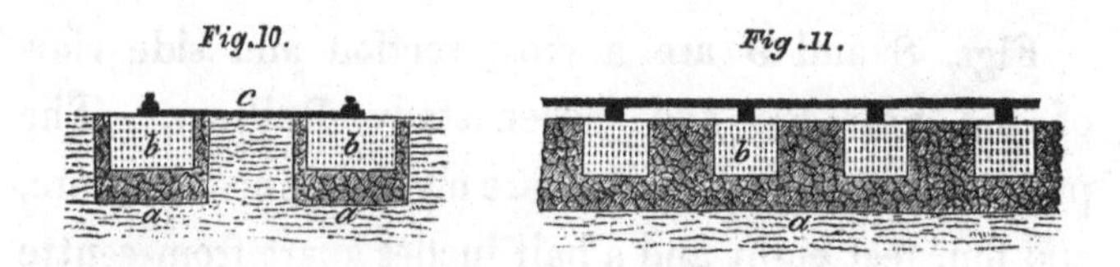

Several experiments have been made on the Columbia Railroad, in Pennsylvania, which is eighty-two miles in length, and is under the management of the State. Part of the road is constructed in accordance with Figs. 10 and 11, which are a transverse section and side view of one of the tracks. The trenches marked *a*, measuring two feet six inches in breadth, and two feet in depth, are excavated in the ground, and filled with broken metal; in these, the stone-blocks, *b*, two feet square, and a foot in thickness, are imbedded at distances of three feet apart, to which the chairs and rails are spiked in the ordinary manner. The rails on each side of the track are connected together by an iron bar, marked *c* in Fig. 10. This attachment is rendered absolutely necessary on many parts of the Columbia Railroad, by the sharpness of the curves, which, at the time when the work was laid out, were not considered so prejudicial on a railway as experience has shewn them to be.

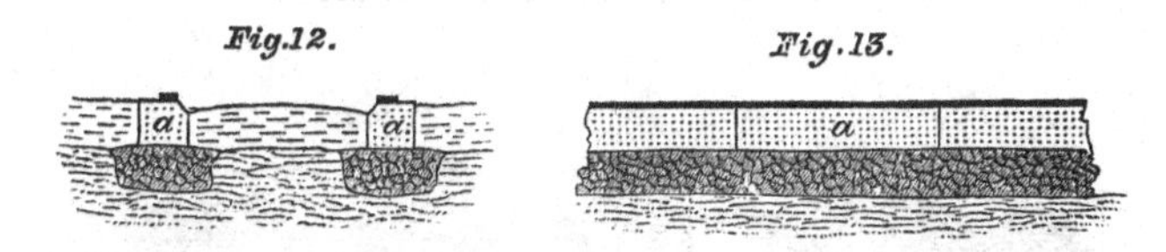

Another plan tried on this line is shewn in Figs. 12 and 13, which are a transverse section and side view. In this arrangement a continuous line of stone curb, one foot square, marked *a*, resting on a stratum of broken stone, is substituted for the isolated stone-blocks, shewn in Figs. 10 and 11. A plate-rail, half an inch thick, and two and a half inches broad, is spiked down to treenails of oak, or locust wood, driven into jumper-holes bored in the stone curb.

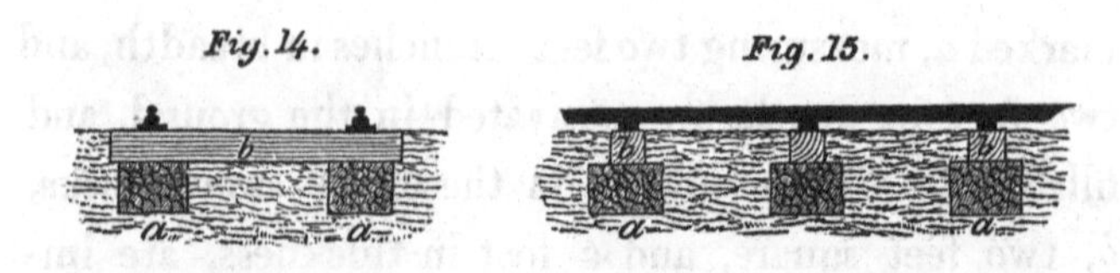

Figs. 14 and 15 represent the construction of the Boston and Providence Railway, which is forty-one miles in length. Pits, measuring eighteen inches square, and one foot in depth, marked *a*, are excavated under each line of rail, at intervals of four feet apart. They are filled with broken stone, and form a foundation for the transverse sleepers, marked *b*, measuring eight inches square, on which the chairs and rails are fixed in the usual manner.

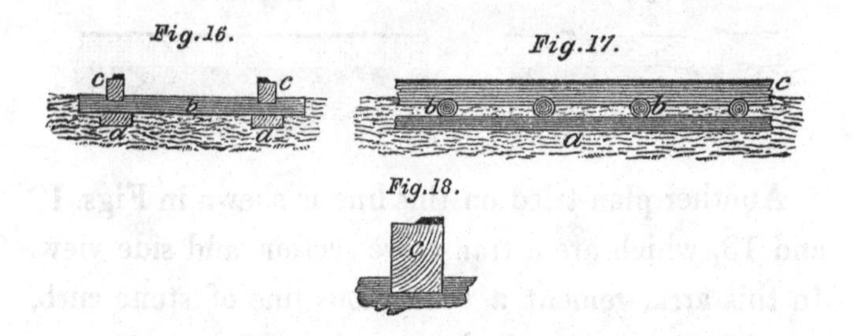

The construction shewn in Figs. 16 and 17, which are a cross section and side view of one of the tracks, is in very general use in America. I met with it on the Philadelphia and Norristown, the New York and Haerlem, and the Buffalo and Niagara railroads; and I believe it has been introduced on many others. It consists of two lines of longitudinal wooden runners, marked *a*, measuring one foot in breadth, and from three to four inches in thickness, bedded on broken stone or gravel. On these runners, transverse sleepers, *b*, are placed, formed of round timber with the bark left on, measuring about six inches in diameter, and squared at the ends, to give them a proper rest. Longitudinal sleepers, *c*, for supporting the rails, are notched into the transverse sleepers, as shewn in the diagram. Fig. 18 is an enlarged view of the plate-rail and longitudinal sleeper used for railways of this construction. The rail is made of wrought-iron, and varies in weight from 10 to 15 lb. per lineal yard. It is fixed down to the sleepers at every fifteen or eighteen inches, by spikes four or five inches in length, the heads of which are countersunk in the rail.

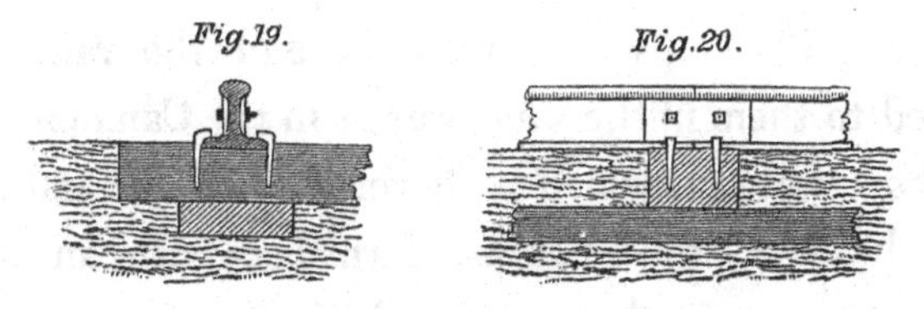

Figs. 19 and 20 are the rails used on the Camden and Amboy Railway, which is sixty-one miles in length. They are parallel edge-rails, and are spiked to transverse sleepers of wood, and, in some places, to wood treenails driven into stone blocks. Their breadth is three and a half inches at the base, and two and a half at the top, and their height is four inches. They are formed in lengths of fifteen feet, and secured at the joints by an iron plate on each side, with two screw-bolts passing through the plates and rails, as shewn in the diagram. On the Philadelphia and Reading Railroad, rails of the same form have been adopted.

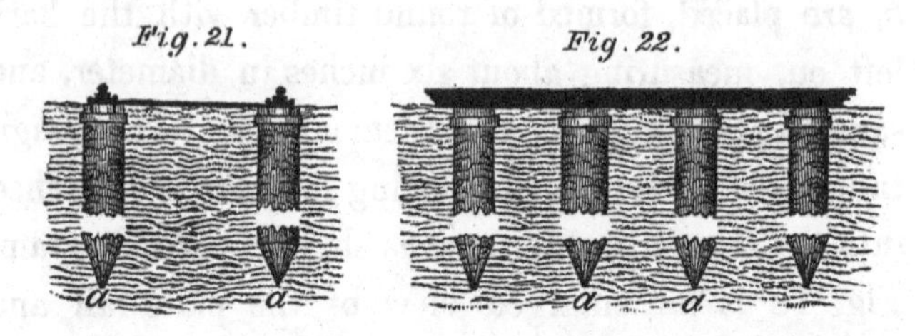

Figs. 21 and 22 shew another construction, which I observed on several of the railroads. It was proposed with a view to counteract the effects of frost. Round piles of timber, marked *a*, about twelve inches in diameter, are driven into the ground as far as they will go, at the distance of three feet apart from centre to centre. The tops are cross-cut, and the rails are spiked to them in the same way as in the Camden and Amboy Railway, which is shewn in Figs. 19 and 20. The heads of the piles are furnished with an iron strap, to prevent them from splitting; and the rails are connected together at every five feet by an iron bar.

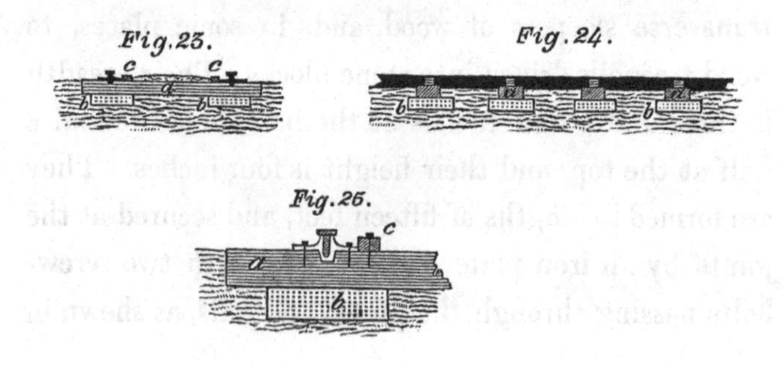

Figs. 23 and 24 are a transverse section and side view of the present structure of the Brooklyn and Jamaica Railroad, on which Mr Douglass, the engineer for that work, has made several experiments. The road, represented in the cut, is exceedingly smooth, and is said to resist the effects of frost very successfully. It consists of transverse sleepers, measuring eight by six inches, marked *a*, supported on slabs of pavement, two feet square, and six inches thick, marked *b*. The wooden runner, marked *c*, is spiked on the inside of the chairs to render them firm. An enlarged view of the rail is shewn at Fig. 25. This rail rests on the *cheeks* or sides of the chair, and not on the bottom, as is generally the case.

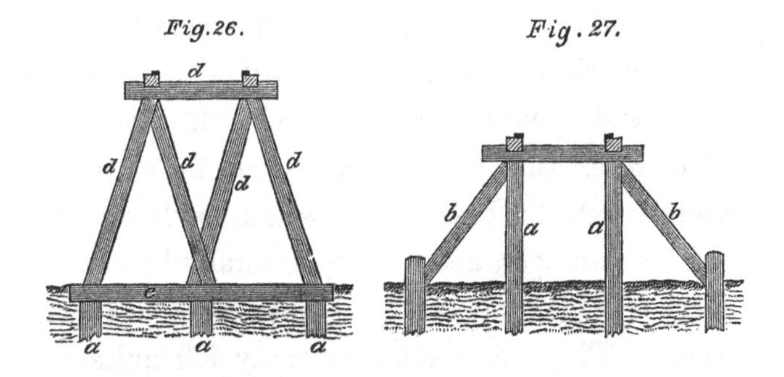

The railroad between Charleston and Augusta, and many others in the southern States, where there is a scarcity of materials for forming embankments, are carried over low lying tracts of marshy ground, elevated on structures of wooden truss-work, such as is shewn in Figs. 26 and 27. The framing in Fig. 27

is used in situations where the level of the rails does not require to be raised more than ten or twelve feet above the surface of the ground. Piles from ten to fifteen inches in diameter, marked *a*, are driven into the ground by a piling engine, and, in places where the soil is soft, their extremities are not pointed but are left square, which makes them less liable to sink under the pressure of the carriages. The struts marked *b* are attached to the tops of the piles, and are also fixed to dwarf piles driven into the ground. Their effect is to prevent lateral motion. Fig. 26 is a truss-work which is used for greater elevations, and is sometimes carried even to the height of fifteen or twenty feet. Piles marked *a* are driven into the ground, and connected by the transverse beam *c*. Above these the superstructure formed of the beams *d* is raised, and upon it, the rails are placed. It is evident, however, that these structures are by no means suitable or safe for bearing the weight of locomotive engines or carriages, and, as may naturally be expected, very serious accidents have occasionally occurred on them. They are besides generally left quite exposed, and in some situations, when they are even so much as twenty feet high, no room is left for pedestrians, who, if overtaken by the engine, can save themselves only by making a leap to the ground.

These varieties of construction were all in use when I visited the United States in 1837, but the American engineers had not at that time come to any definite conclusion as to which of them constituted the best railway. It seemed to be generally admitted, however, that the wooden structures were in most situations more economical than those formed of stone, and were also less liable to be affected by the frost. Structures of wood also possess a great advantage over those of stone, from the much greater ease with which the rails supported by them are kept in repair. Wooden railroads are more elastic, and bend under great weights, while the rigid and unyielding nature of the railroads laid on stone blocks causes the impulses produced by the rapid motion of locomotive carriages, or heavily loaded waggons, over the surface, to be much more severely felt both by the machinery of the engine and by the rails themselves. Experience, both in this country and in America, has shewn the truth of these remarks. On the Liverpool and Manchester Railway, for example, on which a large sum is annually expended in keeping the rails in order, the part of the road which requires least repair is that extending over Chat Moss, where the rails are laid on wooden sleepers, and the weight of passing trains of loaded waggons produces a sensible undulation in the surface of the railway, which at this place actually floats on the moss. These considerations are worthy of attention; and since the introduction of Kyan's patent anti dry-rot preparation, wood is beginning to be more generally employed for the construction of railways in this country. The rails of the Dublin and

Kingstown road are now laid on wood, and it has also been extensively employed on the Great Western Railway now in progress.

The rails used in the United States are of British manufacture. They are often taken to America as ballast; and the Government of the United States having removed the duty from iron imported for the purpose of forming railways, the rails are laid down on the quays of New York nearly at the same cost as in any of the ports of Great Britain. Those of the Brooklyn and Jamaica road, which are in lengths of fifteen feet, and weigh 39 lb. per lineal yard, are of British manufacture, and cost at New York when they were landed, in 1836, L.8 per ton; the cast-iron chairs, which are also of British manufacture, weigh about 15 lb. each, and cost L.9 per ton. There is a great abundance of iron-ore in America, and some of the veins in the neighbourhood of Pittsburg are at present pretty extensively worked; but the Americans know that it would be bad economy to attempt to manufacture rails, so long as those made at Merthyr Tydvil Iron-works, in Wales, can be laid down at their sea-ports at the present small cost. In some of the iron-works which I visited, the workmen were rolling plate-rails, which is the only kind they ever attempt to make; but even these can be got, if not at less cost, at all events of much better quality, from Britain.

The stone blocks in use on some of the railways are made of granite, which, as already noticed, is found in several parts of the United States. Yellow pine is generally employed for the longitudinal sleepers, and cedar, locust, or white-oak, for the transverse sleepers on which the rails rest; cedar, however, if it can be obtained, is generally preferred for the transverse sleepers, because it is not liable to be split by the heat of the sun, and is less affected than perhaps any other timber, by dampness and exposure to the atmosphere. The cedar sleepers used on the Brooklyn and Jamaica Railway, measuring six inches by five, and seven feet in length, notched and in readiness to receive the rails, cost 2s. 3½d. each, laid down at Brooklyn. It is a costly timber, and is not very plentiful in the United States; it has also risen greatly in value since the introduction of railways, for the construction of which it is peculiarly applicable. For all treenails, locust-wood is universally employed.

The American railroads are much more cheaply constructed than those in this country, which is owing chiefly to three causes; *first*, they are exempted from the heavy expenses often incurred in the construction of English railways, by the purchase of land and compensation for damages; *second*, the works are not executed in so substantial and costly a style; and, *third*, wood, which is the principal material used in their construction, is got at a very small cost. The first six miles of the Baltimore and Ohio Railroad, which is formed "in an expensive manner, on a very difficult route," has cost, on an average, about L.12,000 per mile. The railroads in Pennsylvania cost about L.5000 per mile; the Albany and Schenectady Railroad upwards of L.6000 per mile; the Schenectady and Saratoga Railway L.1800 per mile; and the Charleston and Augusta Railroad about the same.* Mr Moncure Robinson, in a report relative to the Philipsburg and Juniata Railroad, states, that the first ten miles of the Danville and Pottsville Railroad, formed for a double track, but on which a single track only was laid, cost on an average L. 4400 per mile, and that the Honesdale and Carbondale Railroad, 16⅓ miles in length, laid with a single track, and executed for a considerable portion of its length on truss-work, is understood, with machinery, to have averaged L.3600 per mile. The average cost of these railways, constructed in different parts of the United States, is L.4942 per mile.

This contrasts strongly with the cost of the railways constructed in this country. The Liverpool and Manchester Railway cost L.30,000 per mile; the Dublin and Kingstown L.40,000; and the railway between Liverpool and London is expected to cost upwards of L.25,000.

The following extract, embodying an estimate from Mr Robinson's Report, will give some idea of the cheapness with which many of the American works are constructed:—

* Facts and suggestions relative to the New York and Albany Railway. New York, 1833.

" The following plan," says Mr Robinson, " is proposed for the superstructure of the Philipsburg and Juniata Railroad.

" Sills of white or post oak, seven feet ten inches long, and twelve inches in diameter, flattened to a width of nine inches, are to be laid across the road at a distance of five feet apart from centre to centre. In notches formed in these sills, rails of white oak or heart pine, five inches wide by nine inches in depth, are to be secured, four feet seven inches apart, measured within the rails. On the inner edges of these rails, plates of rolled iron, two inches wide by half an inch thick, resting at their points of junction on plates of sheet iron, one-twelfth of an inch thick and four and a half inches long, are to be spiked, with five-inch wrought iron spikes. The inner edges of the wooden rails to be trimmed slightly levelling, but flush at the point of contact with the iron rail, and to be adzed down outside the iron to pass off rain-water.

" Such a superstructure as that above described would be entirely adequate to the use of locomotive engines of from fifteen to twenty horses power, constructed without surplus weight, or similar to those now in use on the little Schuylkill Railroad in this state (Pennsylvania), or the Petersburg Railroad in Virginia; and it will be observed that only the sills, which constitute but a very slight item in its cost, are much exposed to the action of those causes which induce decay in timber. It is particularly recommended for the Philipsburg and Juniata Railroad, by the great abundance of good materials along the line of the improvement, for its construction, and the consequent economy with which it may be made.

"The following may be deemed an average estimate of the cost of a mile of superstructure as above described.

	Dollars.
1056 trenches 8 feet long, 12 inches wide, and 14 inches deep, filled with broken stone, at 25 cents each, .	264
Same number of sills, hewn, notched, and embedded, at 50 cents each, ...	528
10,912 lineal feet of rails (allowing $33\frac{1}{3}$ per cent. for waste), at 4 cents per lineal foot, delivered, .	436.48
2112 keys at $2\frac{1}{2}$ cents each, ...	52.80
10,560 lineal feet of plate rails, 2 inches by $\frac{1}{2}$ inch, weight $3\frac{1}{3}$ lb. per foot, $15\frac{71}{100}$ tons, delivered at 50 dollars (L.10) per ton, ...	785.50
1509 lb. of 5-inch spikes, at 9 cents per pound, .	135.81
Sheet iron under ends of rails, ...	30.21
Placing and dressing wood, and spiking down iron rails,	280
Filling between sills with stone, or horse-path, .	180
	2692.80

2692 dollars, or about L.540.

I found it rather difficult to obtain much satisfactory information regarding the expense of upholding the American railways. It is stated in a report made by the Directors of the Boston and Worcester Railroad, that Mr Fessenden, their engineer, to whom I am indebted for much kind attention and valuable information, estimates the annual expenditure for repairing the road, carriages, and engines, and providing fuel and necessary attendance for forty-three and a half miles of railway at L.6829 per annum, which is at the rate of L.157 per mile. The expense of the repairs on the Utica and Schenectady Railroad, which is about seventy-seven miles in length, amounts to L.28,000 per annum, being at the rate of about L.363 per mile. These sums for keeping railroads in repair are exceedingly small, compared with the amount expended in this country for the same purpose. On the Liverpool and Manchester Railway, for example, the expense annually incurred in keeping the engines in a working state and the railway in repair, amounts to upwards of L.30,000 or L.1000 per mile. This difference in the cost arises in a great measure from the comparatively slow speed at which the engines working on the American railways are propelled, which, in the course of my own observation, never exceeded the average rate of fifteen miles per hour. On the State railways, and also on many of those under the management of incorporated companies, fifteen miles an hour is the rate of travelling fixed by the administration of the railway, and this speed is seldom exceeded.

On some of the American railways, where the line is short or the traffic small, horse power is employed, but locomotive engines for transporting goods and passengers are in much more general use. In New York, Brooklyn, Philadelphia, Baltimore, and other places which have lines of railway leading from them, the depôt or station for the locomotive engines is generally placed at the outskirts, but the rails are continued through the streets to the heart of the town, and the carriages are dragged over this part of the line by horses, to avoid the inconvenience and danger attending the passage of locomotive engines through crowded thoroughfares. I travelled by horse power on the Mohawk and Hudson Railway, from Schenectady to Albany, a distance of sixteen miles, and the journey was performed in sixty-five minutes, being at the astonishing rate of fifteen miles an hour. The car by which I was conveyed carried twelve passengers, and was drawn by two horses which ran stages of five miles.

The first locomotive engines used in America were of British manufacture, but several very large workshops have lately been established in the country for the construction of these machines, which are now manufactured in great numbers. The largest locomotive engine-works are those of Mr Baldwin, Mr Norris, Mr Long, and Messrs Grant and Eastrick, all in Philadelphia, and the Lowell Engine-work at Lowell. When I visited the work of Mr Baldwin, to whom I am indebted for much attention and information, I found no less than twelve locomotive carriages in different states of progress, and all of substantial and good workmanship. Those parts of the engine, such as the cylinder, piston, valves, journals, and slides, in which good fitting and fine workmanship are indispensable to the efficient action of the machine, were very highly finished, but the external parts, such as the connecting rods, cranks, framing, and wheels, were left in a much coarser state than in engines of British manufacture. The American engines with their boilers filled, weigh from twelve to fifteen tons, and cost about L.1400 or L.1500, including the tender. This is not much more than the cost of an engine of the same weight in this country. They have six wheels. These are arranged in the following manner, so as to allow the engine to travel on rails having a great curvature ; the driving wheels, which are five feet in diameter, are placed in the posterior part of the engine close to the fire-box, and the fore part of the engine rests on a truck running on four wheels of about two feet six inches in diameter : a series of friction-rollers, arranged in a circular form, is placed on the top of the truck, and in the centre, stands a vertical pivot which works in a socket in the framing of the engine. The whole weight of the cylinders and the fore part of the boiler rests on the friction rollers, and the truck turning on the pivot as a centre, has freedom to describe a small arc of a circle ; so that when the engine is not running upon a perfectly straight road, its wheels adapt themselves to the curvature of the rails, while the relative positions which the body of the engine, the connecting rods, and other parts of the machinery bear to each other, remain unaltered.*

* I believe an attempt was made to apply Avery's Rotatory Engine to propel a locomotive carriage, on one of the American railways, but

From the unprotected state of most of the railways, which are seldom fenced, cattle often stray upon the line, and are run down by the engines, which are in some cases thrown off the rails by the concussion, producing very serious consequences. To obviate this, and render railway travelling more safe, an apparatus called a "guard" has been very generally introduced, a drawing of which is given in Plate XI. Fig. 1. is a side view of a locomotive engine, with the guard attached to it; and Fig. 2. is a plan of the guard and the two front wheels of the engine. The guard consists of a strong framework of wood, marked *a*, fixed to the fore-axle of the locomotive carriage at the point *b*, and supported on two small wheels *c*, about two feet in diameter, which run on the rails about three feet in advance of the engine. The outer extremity of the framework, *d*, is shod with iron slightly bent up, and comes to within an inch of the top of the rails. The upper part of the surface of the guard, as shewn in Fig. 2, is covered with wood, and the lower part with an iron-grating. The apparatus affords a complete protection to the wheels of the engine.

I could not obtain satisfactory information either as to the particulars of the experiment, or the part of the country in which it was made. Avery's engines are, I believe, a good deal used in the northern parts of the United States, for driving small mills. They are generally of from 6 to 12 horses power. In New York I saw three of them at work, one in the Astor Hotel, which was employed to pump water, grind coffee, &c. one in a saw-mill in Attorney Street, and the third working a printing press; these were the only engines constructed on the rotatory principle, which I saw in actual use in the country.

PLATE XI.

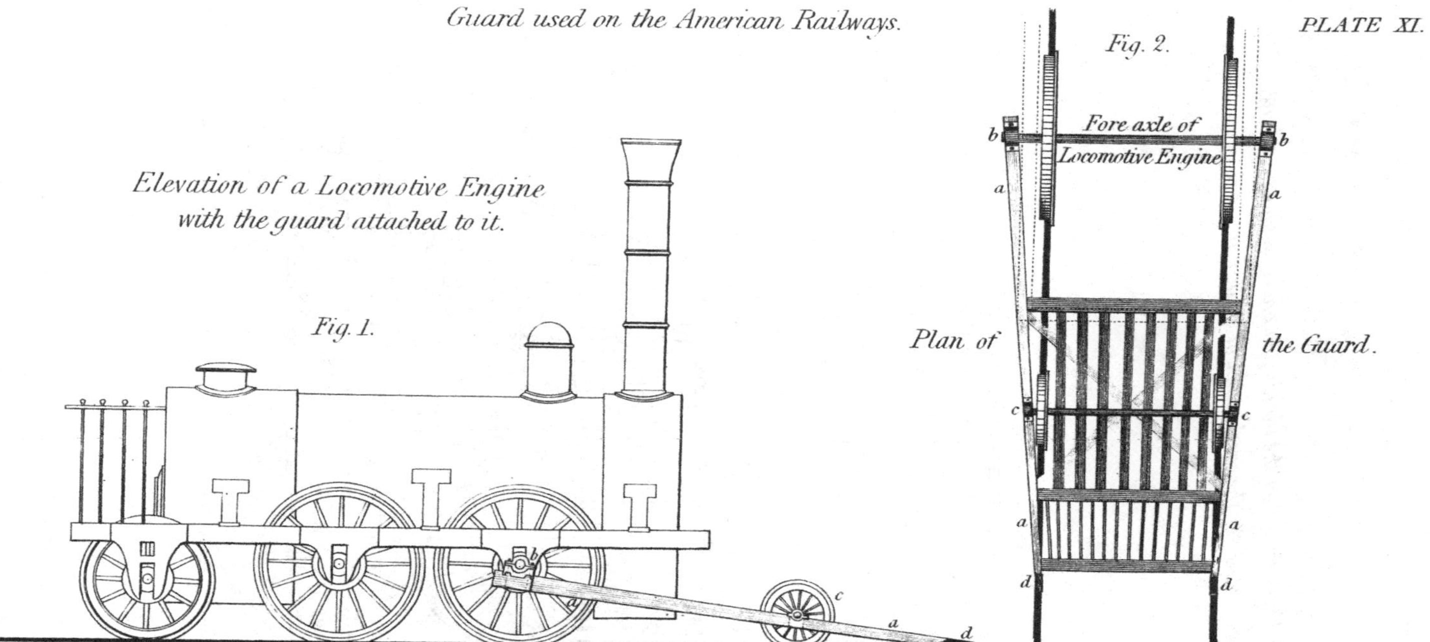

Published by John Weale, 59, High Holborn, 1838.

Stevenson's Sketch of the Civil Engineering of North America.

Geo. Aikman, Sculpt

Locomotive Engine used on the Washington and Baltimore Railway, Constructed for the combustion of Anthracite Coal.

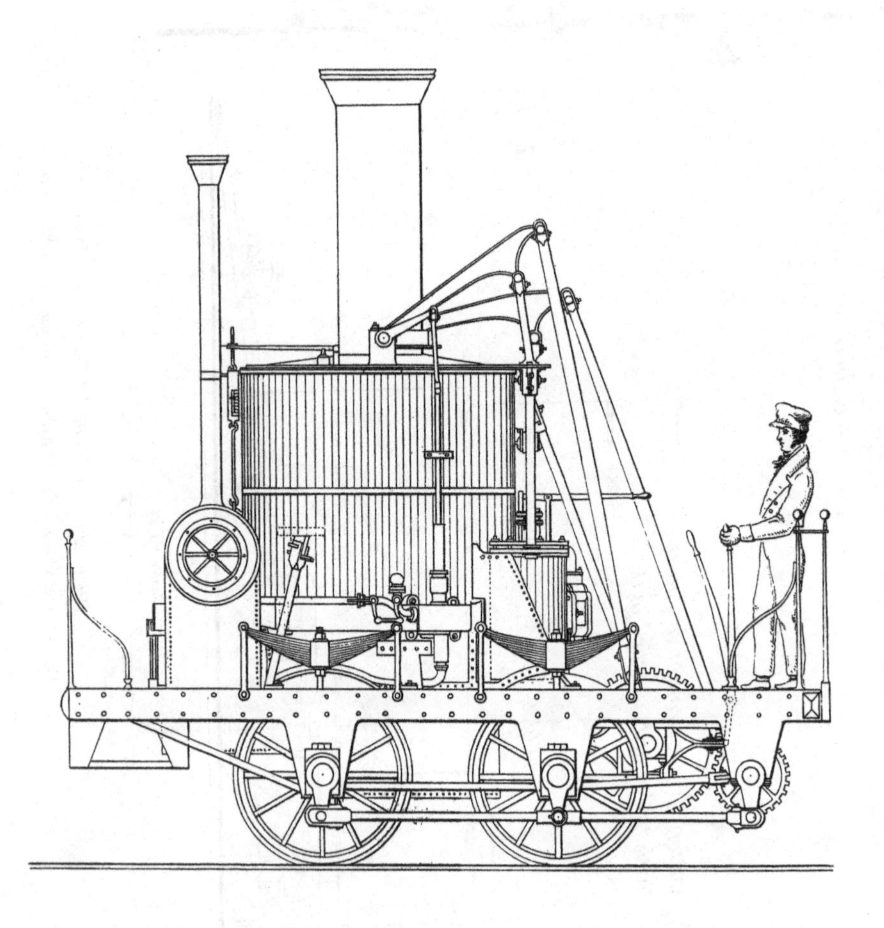

Stevenson's Sketch of the Civil Engineering of North America.

James Andrews. Delt Published by John Weale, 59. High Holborn, 1838. Geo. Aikman, Sculpt

I experienced the good effects of it upon one occasion on the Camden and Amboy Railway. The train in which I travelled, while moving with considerable rapidity, came in contact with a large waggon loaded with firewood, which was literally shivered to atoms by the concussion. The fragments of the broken waggon, and the wood with which it was loaded, were distributed on each side of the railway, but the guard prevented any part of them from falling before the engine-wheels, and thus obviated what might in that case have proved a very serious accident. This apparatus might be introduced with much advantage on the railways in this country, on which accidents, attended with the loss of several lives, have happened from similar causes.

The fuel used on most of the railways is wood, but the sparks vomited out by the chimney are a source of constant annoyance to the passengers, and occasionally set fire to the wooden bridges on the line and the houses in the neighbourhood. Anthracite coal, as formerly noticed, has been tried, but the same difficulties which attend its use in steam-boat furnaces are experienced to an equal extent in locomotive engines. Plate XII. is a drawing of a locomotive carriage used on the Baltimore and Washington Railway, constructed by Gillingham and Winans at Baltimore, which is adapted to the use of anthracite coal. It has vertical cylinders, with a vertical tubular boiler, and weighs about eight tons.

In situations where the summit level of a railway cannot be attained by an ascent sufficiently gentle for the employment of locomotive engines, or where the formation of such inclinations, though perfectly practicable, would be attended with an unreasonably large outlay, transit is generally effected by means of inclined planes, worked by stationary engines. This system has been introduced on the Portage Railway over the Alleghany Mountains in America, on a more extensive scale than in any other part of the world. The Portage, or Alleghany Railway, forms one of the links of the great Pennsylvania canal and railroad communication from Philadelphia to Pittsburg,—a work of so difficult and vast a nature, and so peculiar, both as regards its situation and details, that it cannot fail to be interesting to every engineer, and I shall, therefore, state at some length the facts which I have been able to collect regarding it.

This communication consists of four great divisions, the Columbia Railroad, the Eastern Division of the Pennsylvania Canal, the Portage or Alleghany Railroad, and the Western Division of the Pennsylvania Canal. These works form a continuous line of communication from Philadelphia on the Schuylkill to Pittsburg on the Ohio, a distance of no less than 395 miles.

Commencing at Philadelphia, the first Division of this stupendous work is the Philadelphia and Columbia Railroad, which was opened in the year 1834. It is eighty-two miles in length, and was executed at a cost of about L.666,025, being at the rate of L.8122 per mile. There are several viaducts of considerable extent on this railway, and two inclined planes worked by stationary engines. One of these inclined planes is at the Philadelphia end of the line. It rises at the rate of one in 14.6 for 2714 feet, overcoming an elevation of 185 feet. The other plane which is at Columbia rises at the rate of one in 21.2 for a distance of 1914 feet, and overcomes an elevation of 90 feet. A very large sum is incurred in upholding the inclined planes, and surveys have lately been made with a view to avoid them. The cost of maintaining the stationary power, and superintendence of the Philadelphia inclined plane, is said to be about L.8000 per annum, and that of the Columbia plane about L.3498 per annum. Locomotive engines are used between the tops of the inclined planes. The steepest gradient on that part of the line is at the rate of one in 117; but the curves are numerous, and many of them very sharp, the minimum radius being so small as 350 feet. This line of railway was surveyed and laid out before the application of locomotive power to railway conveyance had attained its present advanced state,—at a period when sharp curves and steep gradients were not considered so detrimental to the success of railways as experience has since shewn them to be.

The passenger carriages on the Columbia Railroad

are extremely large and commodious. They are seated for sixty passengers, and are made so high in the roof, that the tallest person may stand upright in them without inconvenience. There is a passage between the seats, extending from end to end, with a door at both extremities; and the coupling of the carriages is so arranged, that the passengers may walk from end to end of a whole train without obstruction. In winter they are heated by stoves. The body of each of these carriages measures from fifty to sixty feet in length, and is supported on two four-wheeled trucks, furnished with friction-rollers, and moving on a vertical pivot, in the manner formerly alluded to in describing the construction of the locomotive engines. The flooring of the carriages is laid on longitudinal beams of wood, strengthened with suspension-rods of iron.

At the termination of the railway at Columbia, is the commencement of the Eastern Division of the Pennsylvania Canal, which extends to Hollidaysburg, a town situate at the foot of the Alleghany Mountains. This canal is rather more than 172 miles in length, and was executed at an expense of L.918,829, being at the rate of L.5342 per mile. There are 33 aqueducts, and 111 locks on the line, and the whole height of lockage is 585.8 feet. A considerable part of this canal is slackwater navigation, formed by damming the streams of the Juniata, and Susquehanna. The canal crosses the Susquehanna at its junction with the Juniata, at which point it attains a considerable breadth. A dam has been erected in the Susquehanna at this place, and the boats are dragged across the river by horses, which walk on a tow-path attached to the outside of a wooden bridge, at a level of about thirty feet above the surface of the water. I regret that I passed through this part of the canal after sunset, and had only a very superficial view of the works at this place, which are of an extensive and curious nature.

Hollidaysburg is the western termination of the Eastern Division of the Pennsylvania Canal. The town stands at the base of the Alleghany Mountains, which extend in a south-westerly direction, from New Brunswick, to the State of Alabama, a distance of upwards of 1100 miles, presenting a formidable barrier to communication between the eastern and western parts of the United States. The breadth of the Alleghany range varies from a hundred to a hundred and fifty miles, but the peaks of the mountains do not attain a greater height than 4000 feet above the medium level of the sea. They rise with a gentle slope, and are thickly wooded to their summits. "The Alleghany Mountains present what must be considered their scarp or steepest side to the east, where granite, gneiss, and other primitive rocks are seen. Upon these repose first, a thin formation of transition rocks dipping to the westward, and next a series of secondary rocks, including a very extensive coal formation."* The National Road, which has

* Encyclopædia Brit., article America.

already been noticed, was the first line of communication formed by the Americans over this range ; and in the year 1831, an Act was passed for connecting the Eastern and Western Divisions of the Pennsylvania Canal by means of a railroad. This important and arduous work, which cost about L. 526,871, was commenced within the same year in which the Act for its construction was granted, and the first train passed over it on the 26th of November 1833, but it was not till the year 1835, that both the tracks were completed, and the railway came into full operation.

The railway crosses the mountains by a pass called "Blair's Gap," where it attains its summit level, which is elevated 2326 feet above the mean level of the Atlantic Ocean. Mr Robinson surveyed a line of railway from Philipsburg to the river Juniata, which is intended to cross the Alleghany Mountains by the pass called "Emigh's Gap." The summit level of this line is stated, in a report by the directors, to be 292 feet lower than that of the Portage railway.

The preliminary operation of clearing a track for the passage of the railway from a hundred to a hundred and fifty feet in breadth, through the thick pine forests with which the mountains are clad, was one in which no small difficulties were encountered. This operation, which is called *grubbing*, is little known in the practice of engineering in this country, and is estimated by the American engineers, in their various railway and canal reports, at from L.40 to L.80 per mile, according to the size and quantity of the timber to be removed ; an estimate which, from the appearance of American forests, I should think must in many instances be much too low. The timber removed from the line of the Alleghany railway was chiefly spruce and hemlock pine of very large growth. I passed over the Alleghany Mountains on the 11th of May, at which time the trees were thickly covered with foliage, and formed a wall on each side of the railway, which completely intercepted the view of the surrounding country during the greater part of the journey. An extensive view was occasionally obtained from the tops of the inclined planes, when nothing but a dense black forest was visible, stretching in all directions as far as the eye could reach.

The line is laid with a double track, or four single lines of rails, and is twenty-five feet in breadth. For a considerable distance the railway is formed by side-cutting along steep sloping ground, composed of clay-slate, bituminous coal and clay, part of the breadth of the road being obtained by cutting into the hill, and part by raising embankments protected by retaining walls of masonry. The railway is consequently liable to be deluged, or even entirely swept away, by mountain torrents, and the thorough drainage of its surface has been attended with great expense and difficulty. The retaining walls by which the embankments are supported, are in some places not less than a hundred feet in height; they are built of dry-stone masonry, and have a batter of about one-half to one, or six inches horizontal to twelve inches perpendicular. There are no parapet or fence walls on the railway, and on many parts of the line, especially at the tops of several of the inclined planes, the trains pass within three feet of precipitous rocky faces, several hundred feet high, from which the large trees growing in the ravines below, almost resemble brushwood. One hundred and fifty-three drains and culverts, and four viaducts, have been built on the railway. One of the viaducts crosses the river Conemaugh at an elevation of seventy feet above the surface of the water. There is also a tunnel on the line 900 feet in length, twenty feet in breadth, and nineteen feet in height.

The inclined planes are, however, the most remarkable works which occur on this line. The railway extends from Hollidaysburg on the eastern base, to Johnstown on the western base of the Alleghany Mountains, a distance of thirty-six miles; and the total rise and fall on the whole length of the line is 2571.19 feet. Of this height, 2007.02 feet are overcome by means of ten inclined planes, and 564.17 feet by the slight inclinations given to the parts of the railway which extend between these planes. The distance from Hollidaysburg to the summit-level is about ten miles, and the height is 1398·31 feet. The distance from Johnstown to the same point is about twenty-six miles, and the height 1172.88 feet. The height of the summit-level of the railway above the mean level of the Atlantic is 2326 feet.

The following are the lengths, gradients, and elevations overcome by the several inclined planes, five of which are placed on each side of the summit-level :—

No of Plane.	Length in Feet.	Gradient.	Height overcome.
Plane No. 1.	1607.74	One in 10.71	150 feet.
... 2.	1760.43	... 13.29	132.40 ...
... 3.	1480.25	... 11.34	130.50 ...
... 4.	2195.94	... 11.68	187.86 ...
... 5.	2628.60	... 13.03	201.64 ...
... 6.	2713.85	... 10.18	266.50 ...
... 7.	2655.01	... 10.19	260.50 ...
... 8.	3116.92	... 10.13	307.60 ...
... 9.	2720.80	... 14.35	189.50 ...
... 10.	2295.61	... 12.71	180.52 ...

The following table shews the length of each section of the railway between the inclined planes, and the elevation overcome on it :—

	Length in miles.	Gradient.	Height overcome.
From Johnstown to foot of plane No. 1, . . .	4.13	1 in 214.92	101.46
— head of plane No. 1 to foot of plane No. 2,	13.06	— 363.73	189.58
— do. No. 2 to do. No. 3,	1.43	— 477.87	15.80
— do. No. 3 to do. No. 4,	1.90	— 533.61	18.80
— do. No. 4 to do. No. 5,	2.56	— 523.90	25.80
— do. No. 5 to head of plane No. 6,	1.62	— 449.24	19.04
— foot of plane No. 6 to head of plane No. 7,	0.15	level	
— do. No. 7 to do. No. 8,	0.61	— 596.44	5.40
— do. No. 8 to do. No. 9,	1.18	— 519.20	12.00
— do. No. 9 to do. No. 10,	1.70	— 303.44	29.58
— do. No. 10 to Hollidaysburg, .	3.72	— 133.88	146.71

The machinery by which the inclined planes are worked consists of an endless rope passing round horizontal grooved wheels placed at the head and foot of the planes, which are furnished with a powerful break for retarding the descent of the trains. The ropes were originally made 7½ inches in circumference, but they have lately been increased to 8 inches, to prevent a tendency which they formerly had to slip in the grooved wheels, occasioned by their circumference being too small for the size of the groove or hollow in the wheel. Two stationary engines of twenty-five horses' power each are placed at the head of the inclined planes, one of which is in constant use in giving motion to the horizontal wheels round which the rope moves while the trains are passing the inclined planes. Two engines have been placed at each station, that the traffic of the railway may not be stopped should any accident occur to the machinery of that which is in operation; and they are used alternately for a week at a time. Water for supplying the boilers has been conveyed at a great expense to many of the stations in wooden pipes upwards of a mile in length.

The planes are laid with a double track of rails, and an ascending and a descending train are always attached to the rope at the same time. Many experiments have been made to procure an efficient safety car to prevent the trains from running to the foot of the inclined plane, in the event of the fixtures by which they are attached to the endless rope giving way. Several of these safety-cars are in use, and are found to be a great security. The trains are attached to the endless rope simply by two ropes of smaller size made fast to the couplings of the first and last waggons of the train, and to the endless rope by a hitch or knot, formed so as to prevent it from slipping.

Locomotive engines are used on the parts of the road between the inclined planes.

The following extract from the Report of the Pennsylvania Canal Commissioners for 1836, affords the best proof of the traffic which the road is capable of sustaining.

" The Portage Railway, however complicated in its operations, and limited in capacity by inclined planes, as canals are by locks, is nevertheless adequate to the transaction of a vast amount of business. Occupying as it does, a nearly central position on the main line between Columbia and Pittsburg, the capacity of the planes ought to be equal to that of the canal locks on those Divisions. Many suppose the planes fall very far short of that limit, and that their full capacity is nearly reached.

" It is, however, due to our commercial interest and the public at large, to state that the maximum of that limit is very far from being attained. The length of the longest plane is about 3000 feet ; the time occupied in moving up or down it is five minutes ; the time occupied in attaching is two and a half minutes, making seven and a half minutes, or eight drafts per hour of three loaded cars, carrying three tons each, making twenty-four cars, or seventy-two tons per hour.

"It will be observed by the Report of the Superintendent, that the number of cars weighed at Hollidaysburg and transported from east to west, from April 1st to October 31st, is 14,300, making a transit of a number not exceeding a hundred per day; but, instead of this number, when the trade demands it, twenty-four cars can be passed up and the same number down the longest plane in each hour, making two hundred and eighty-eight cars in the day of twelve hours, or five hundred and seventy-six in one direction in twenty-four hours; this can be accomplished by using the road day and night, by means of a double set of hands. This is the true limit of the capacity of the road."

From the same report it appears, that from the 1st of April to the 31st of October, the time during which the railway was open in the year 1836, 19,171 passengers were conveyed along the line; and the following is a statement of the merchandise weighed at the weigh-scales at Hollidaysburg during the same period, amounting to 37,081 tons, conveyed in 14,300 waggons.

Months.	Merchandise.	Iron.	Coal.	Lumber.	Number of Cars.
April,	7,192,310	1,863,170	673,060	315,435	1,323
May,	13,262,218	1,654,495	2,335,390	258,940	3,208
June,	5,146,415	3,389,160	2,384,735	367,045	1,947
July,	4,724,830	1,843,760	1,019,070	63,310	1,335
August,	8,124,370	2,076,820	2,094,300	347,950	2,183
Sept.	7,132,345	2,063,645	3,645,660	86,620	2,324
October,	5,899,050	1,938,710	2,899,730	260,140	1,980
	51,481,538	14,829,760	15,051,945	1,699,440	14,300

The travelling on this railway is very slow. The train by which I was conveyed left Hollidaysburg at nine in the morning, reached the summit at twelve, where it stopped an hour for dinner, and arrived at Johnstown at five in the evening, seven hours having been occupied in travelling thirty-six miles, being only at the rate of about five miles an hour. Much time is lost in ascending and descending the inclined planes, and an hour is generally spent for dinner at an inn on the summit, which is the only house unconnected with the works which is met with on the whole journey.

The fourth division of this grand work is the Western Division of the Pennsylvania Canal, which extends from the termination of the Portage Railway at Johnstown to Pittsburg. It has 64 locks, 16 aqueducts, 64 culverts, 152 bridges, and a tunnel upwards of 1000 feet in length. This canal traverses the valleys of the Conemaugh, Kiskiminetas, and Alleghany Rivers, measures 105 miles in length, and cost L.560,000, being at the rate of L.5333 per mile.

The whole distance of the Pennsylvania canal and railroad communication, extending from Philadelphia to Pittsburg, is 395 miles. I travelled this distance in ninety-one hours, exclusively of the time lost in stopping at Columbia, Harrisburg, and other places of interest on the route. The average rate of travelling was therefore 4.34 miles per hour. One hundred and eighteen miles of this extraordinary journey were performed on railroads, and the remaining 277 miles on canals. The charge made for each passenger conveyed the whole distance was L.3, being at the rate of nearly 2d. per mile.

There is only one railway in the British dominions in North America. It extends from St John's on Lake Champlain to the village of La Prairie on the St Lawrence, and was made by a company of private individuals, called the Champlain and St Lawrence Railroad Company, who obtained their act of Parliament in 1832. The railway is sixteen miles in length, and consists of plate-rails laid on wooden sleepers. There are no works of importance connected with it, as the line passes through an extensive prairie of low lying level land very favourable for its construction. Two locomotive engines are used on the railway, one of which was made in England and the other in the United States.

TABLE of the Principal Railways in operation in the United States in 1837.

Name.	Course.	When opened.	Length in Miles.	Whole length in each State.
	MAINE.			
Bangor and Orono, .	From Bangor to Orona, .	1836	10	10
	MASSACHUSETTS.			
Quincy, . . .	Quincy Quarries to Neponset River, ...	1827	4	
Boston and Lowell, .	Boston to Lowell, . .	1835	26	
Boston and Providence,	Boston to Providence, . .	1835	41	
Dedham Branch, . .	Boston and Providence Railroad to Dedham, .	1835	2	
Boston and Worcester,	Boston to Worcester, . .	1835	44	
Andover and Wilmington,	Andover to the Boston and Lowell Railroad, .	1836	7¾	
Taunton Branch, . .	Boston and Providence Railroad and Taunton, .	1836	11	
Andover and Haverhill,	Andover to Haverhill, . .	1837	10	
Providence & Stonington,	Providence to Stonington,	1837	47	192¾
	NEW YORK.			
Mohawk and Hudson,	Between the Rivers Mohawk and Hudson, . . .	1832	16	
Saratoga & Schenectady,	Saratoga to Schenectady, .	1832	22	
Rochester, . . .	Rochester to Carthage, .	1833	3	
Ithica and Oswego, .	Ithica to Oswego, . . .	1834	29	
Rensselaer and Saratoga,	Troy to Ballston, . . .	1835	24½	
Utica and Schenectady,	Utica to Schenectady, .	1836	77	
Buffalo and Niagara,	Buffalo to Niagara Falls, .	1837	21	
Haerlem, . . .	New York to Haerlem, .	1837	7	
Lockport and Niagara,	Lockport to Niagara Falls,	1837	24	
Brooklyn and Jamaica,	Brooklyn to Jamaica, . .	1837	12	235½
	NEW JERSEY.			
Camden and Amboy, .	Camden to Amboy, . .	1832	61	
Paterson, . . .	Paterson to Jersey, . .	1834	16½	
New Jersey, . . .	Jersey City to New Brunswick,	1836	31	108½
	PENNSYLVANIA.			
Columbia, . . .	Philadelphia to Columbia, .	...	82	
Alleghany, . . .	Hollidaysburg to Johnstown, over the Alleghany Mts.	...	36	
Mauch Chunk, . .	Mauch Chunk to the Coal-mines, ...	1828	5	
Room Run, . . .	Mauch Chunk to Coal-mines,	...	5½	
Mount Carbon, . .	Mount Carbon to Coal-mines,	1830	7½	
Schuylkill Valley, .	Port Carbon to Tuscarora, with numerous branches,	...	30	
Schuylkill, . . .	. . . .	...	13	
Mill Creek, . . .	Port Carbon to Mill Creek, .	...	7	
Minehill and Schuylkill,	. . . .	...	20	
Pine-grove, . . .	Pine-grove to Coal-mines, .	...	4	
Little Schuylkill, . .	Port Clinton to Tamaqua, .	1831	23	
Lackawaxen, . .	Lackawaxen Canal to the River Lackawaxen, .	...	16½	
	Carry forward,		249½	546¾

Name.	Course.	When opened.	Length in Miles.	Whole length in each State.
	Brought forward,	..	249½	546¾
	PENNSYLVANIA *continued.*			
Westchester,	Westchester to Columbia Railroad,	1832	9	
Philadelphia and Trenton,	Philadelphia to Trenton,	1833	26¼	
Philadelphia & Norristown,	Philadelphia to Norristown,	1837	19	
Central Railway,	Pottsville to Danville,	...	51½	
				355¼
	DELAWARE.			
Newcastle & Frenchtown,	Newcastle to Frenchtown,	1832	16	
				16
	MARYLAND.			
Baltimore and Ohio,	Completed to Harper's Ferry, with branches,	1835	86	
Winchester,	Harper's Ferry to Winchester,	...	30	
Baltimore & Port-Deposit,	Baltimore to Port-Deposit,	...	34¼	
Baltimore & Washington,	Baltimore to Washington,	1835	40	
Baltimore & Susquehanna,	Baltimore to York,	1837	59½	
				249¾
	VIRGINIA.			
Chesterfield,	Richmond to Chesterfield Coal-mines,	...	13	
Petersburg and Roanoke,	Petersburg to Blakely on the Roanoke,	...	59	
Winchester and Potomac,	Winchester to Harper's Ferry,	...	30	
Portsmouth and Roanoke,	Portsmouth to Weldon,	...	77½	
Richmond, Fredricksburg, and Potomac,	Richmond to Fredricksburg,	...	58	
Manchester,	Richmond to Coal-mines,	...	13	
				250½
	SOUTH CAROLINA.			
S. Carolina Railroad,	Charleston to Hamburg on the Savannah,	1833	136	
				136
	GEORGIA.			
Alatamaba & Brunswick,	Alatamaba to Brunswick,	...	12	
				12
	ALABAMA.			
Tuscumbia and Decatur,	Mussel-Shoals, Tenessee River,	...	46	
				46
	LOUISIANA.			
Pontchartrain,	New Orleans to Lake Pontchartrain,	1831	5	
Carolton,	New Orleans to Carolton,	...	6	
				11
	KENTUCKY.			
Lexington and Ohio,	Lexington to Frankfort,	...	29	
				29
	Total length in miles,			1652¼

LIST OF THE PRINCIPAL RAILWAY WORKS NOW IN PROGRESS IN THE UNITED STATES.

NAME.	COURSE.	Length in Miles.
	NEW HAMPSHIRE.	
Nashua and Lowell, .	Nashua to Lowell, ...	15
	MASSACHUSETTS.	
Eastern Railroad, .	Boston to Portsmouth, ...	50
Worcester and Norwich,	Worcester to Norwich, ...	58
Western Railway, .	Worcester to Springfield, ...	35
	CONNECTICUT.	
Hartford and Newhaven,	Hartford to Newhaven, ...	35
	NEW YORK.	
Auburn and Syracuse,	Auburn to Syracuse, ...	26
Catskill and Canajoharie,	Catskill to Canajoharie, ...	68
Hudson and Berkshire,	Hudson to the Boundary of Massachusetts,	30
Long Island, . .	Jamaica to Greenport, ...	50
New York and Erie, .	New York to Lake Erie, ...	505
Saratoga and Washington,	Saratoga to Whitehall, ...	41
Tonawanta, . . .	Rochester to Attica, ...	45
	NEW JERSEY.	
Elisabethtown & Belvidere,	Elisabethtown to Belvidere, . .	60
Burlington & Mount Holly,	Burlington to Mount Holly, . .	7
Morris and Essex, .	Morristown to Newark, ...	20
	PENNSYLVANIA.	
Philadelphia and Reading,	Philadelphia to Reading, ...	40½
Oxford, ...	Columbia Railroad to Port Deposit, .	38
Philadelphia and Baltimore,	Philadelphia to Baltimore, ...	93
Tioga, ...	Chemung Canal to Tioga Coal-Mines,	40
	VIRGINIA.	
Greensvill and Roanoke,		18
	S. CAROLINA.	
Augusta and Athens, .	Augusta to Athens, ...	100
Charleston and Cincinnatti,	Charleston to Cincinnati, ...	500
	GEORGIA.	
Macon and Forsyth, .	Macon to Forsyth, ...	25
Central Railroad, . .	Savannah to Macon, ...	200
	ALABAMA.	
Montgomery and Chattahoochee, . . .		90
	MISSISSIPPI.	
Mississippi Railroad, .	Natchez to Canton, ...	150
	KENTUCKY.	
Frankfort and Louisville,	Frankfort to Louisville, ...	50
Bowling Green & Barren River, . . .	Bowling-Green to Barren River, . .	1½
	OHIO.	
Mud River and Lake Erie,	Dayton to Sandusky, ...	153
Sandusky and Monroeville,	Sandusky to Monroeville, ...	16
	MICHIGAN.	
Detroit and St Joseph, .	Detroit to the River St Joseph, . .	200
	Total length, .	2760

CHAPTER X.

WATER-WORKS.

Fairmount Water-works at Philadelphia—Construction of the Dam over the River Schuylkill—Pumps and Water-wheels—Reservoirs, &c.—The Water-works of Richmond in Virginia—Pittsburg—Montreal—Cincinnatti—Albany—Troy—Wells for supplying New York and Boston—Plan for improving the supply of Water for New York, &c.

THE Fairmount Water-works are situate on the east bank of the river Schuylkill, about one mile and a half from the town of Philadelphia. They are remarkable for their efficiency and simplicity, as well as their great extent, being the largest water-works in North America. They were commenced in 1819, and were in a working state in 1822. According to the Water Company's report for the year 1836, the whole sum expended in their execution, up to that date, was L.276,206.*

The water of the river Schuylkill, with which the town of Philadelphia is supplied, is raised by water-

* Annual Reports of the Watering Committee to the Select and Common Councils of the City of Philadelphia.

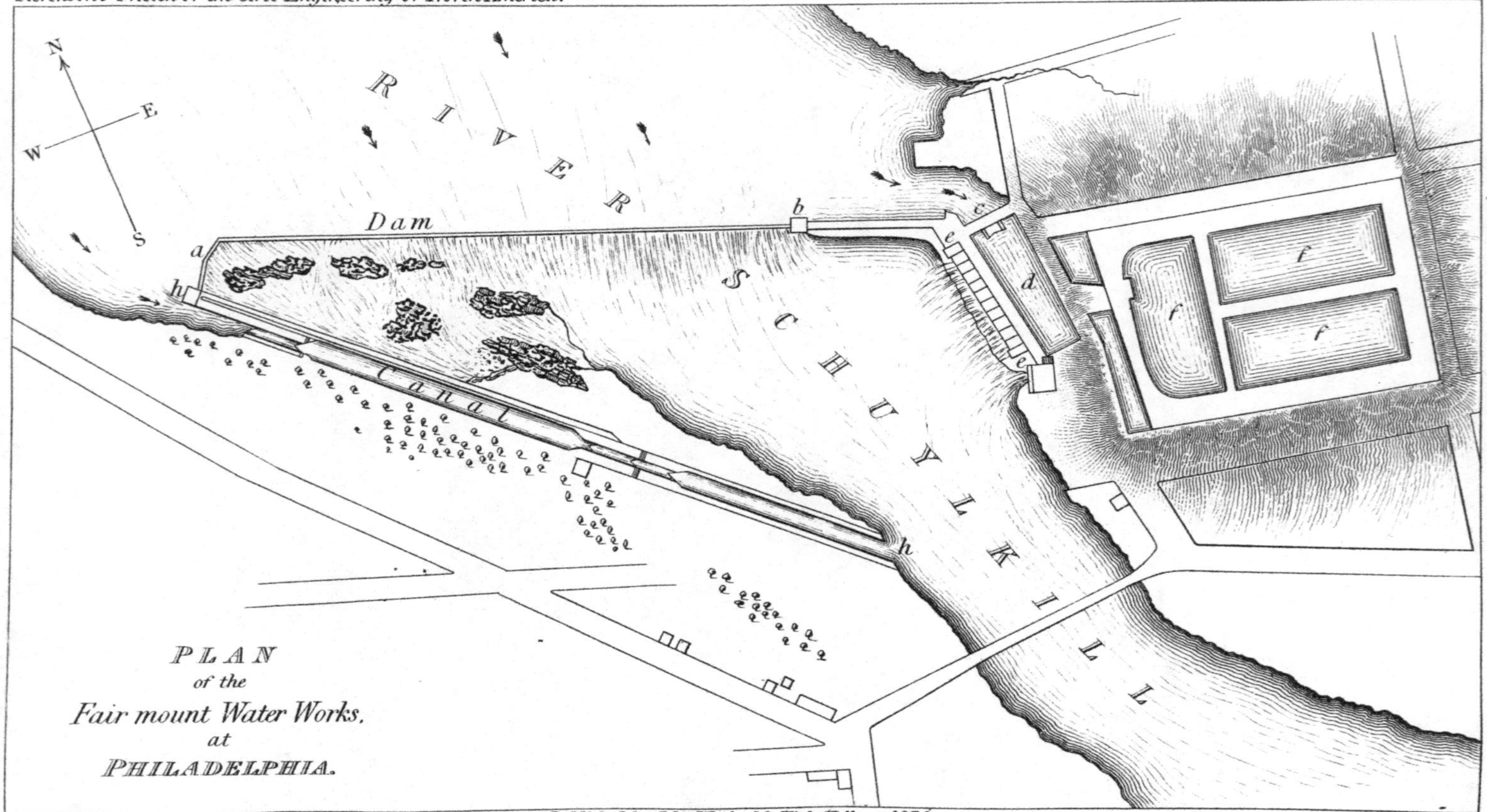

PLAN
of the
Fair mount Water Works,
at
PHILADELPHIA.

Thomas Stevenson, Delt.

Published by John Weale, 59, High Holborn, 1838.

Geo. Aikman, Sculp.

power into four large reservoirs, placed on a rocky eminence near the bank of the river; and after passing through gravel filter-beds, it is conveyed in two large mains to the outskirts of the town, and thence led into the various streets by smaller mains and branch-pipes.

Plate XIII. is a ground plan of the water-works, including part of the river Schuylkill and the adjoining country. Letters *a b c* represent a dam which has been thrown across the river in order to obtain a fall of water for driving the water-wheels. Letter *d* is the mill-race, *e e* the buildings in which the water-wheels and force-pumps are placed, and *f f f* are the filters and reservoirs for the reception of the water.

The erection of the dam across the river was the first and most arduous part of this work. It measures about sixteen hundred feet in length from bank to bank, and creates a stagnation in the flow of the stream, which extends about six miles up the river. The greatest depth of water in the line of the dam at low water of spring tides is twenty-four feet, and the rise of tide is six feet. From *c* to *b* the bottom of the river consists of rock covered with a deposit of mud about eleven feet in depth, and from *b* to *a* the bottom is entirely composed of bare rock, part of which, at the western side of the river, is exposed during low water, as shewn in the plate. The line of the dam forms an angle of about 45 degrees with the direction of the stream. In this way a large overfall is formed for the water, and its perpendicular rise above the top of the dam, when the river is in a flooded state, is not so great as it would have been had the dam been placed at right angles to the stream. By adopting this direction the strength of the structure is also considerably increased, for the mass of the dam opposed to any given section of the stream is greater directly as the cosine, or inversely, as the sine of the angle formed by the line of the dam and the direction of the stream inpinging on it.

The part of the dam which was first formed is that which is founded on the mud bottom extending from *c* to *b*. It consists of a large mound composed of rubble stones and earth thrown into the river. It measures 270 feet in length, 150 feet in breadth at the base, and 12 feet at the top, and its upper slope or face, which is exposed to the wash of the river, is cased with rough pitching formed of large stones. The termination of the dam at the point *b*, is protected by a cut-stone pier, measuring twenty-eight feet by twenty-three feet, which is founded on rock, and built in water twenty-eight feet in depth.

The part extending from *b* to *a* is the overfall dam. It measures 1204 feet in length, and is founded on a rocky bottom, which rises pretty regularly from *b*, where there is a depth of twenty-four feet during the lowest tides, towards *a*, where the rock is uncovered at low water.

The current of the river being strong, it was found

Fig. 1.

Fig. 2.

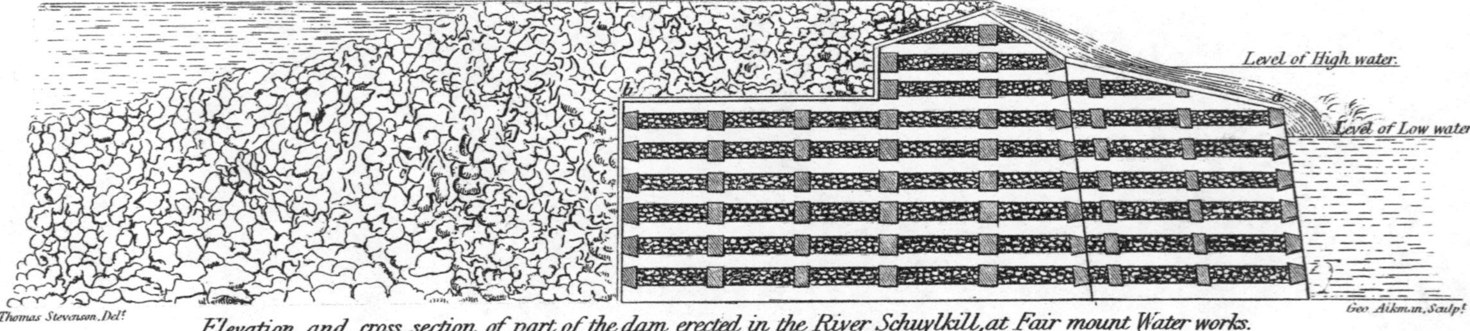

Thomas Stevenson, Delt.

Geo Aikman, Sculpt.

Elevation and cross section of part of the dam erected in the River Schuylkill, at Fair mount Water works.

Published by John Weale, 59. High Holborn, 1838.

impossible to form this part of the dam by constructing a mound of rubble on the rocky bottom, according to the plan followed in founding the first part of the structure, on a bottom composed of mud. The expedient resorted to for retaining the stones on the shelving rock, was extremely ingenious, and has proved very effective.

The overfall dam consists of a strong wooden framework or crib, which was formed in separate compartments, and sunk in small portions in the line of the dam, by filling it with stones. Plate XIV. is a drawing of the dam, in which Fig. 1 is an elevation of a part of its lower front or face, and Fig. 2 is a cross section. These views shew the wooden frames or cribs of which the dam is composed, and also the rubble-stone hearting which prevents them from floating. The cribs are formed of logs of wood, measuring eighteen by twenty inches, connected together by strong dovetailing, notched three inches deep, in the manner shewn in the drawing. The size of the wooden framework, measured in the direction of the stream, is seventy-two feet, and the separate compartments of which it was formed measured twenty feet in breadth. The part of the dam over which the water flows marked *a a*, and also the posterior part of it, *a b*, are covered with planking six inches in thickness. In forming the dam, the cribs were floated one after another to the site which they were to occupy, and large stones being thrown into them, they gradually sank, until at last they rested on the bottom of the river. The upper parts of the several cribs, or those portions of them which stood above the level of low water, were then firmly connected together, so as to form one continuous frame-work, behind which a large mass of rubble hearting and earth was placed, in the manner shewn in the drawing, to give the whole structure weight and stability, and to prevent leakage.

This mode of forming dams is very generally practised in America in forming lines of slackwater navigation, and has been found to stand remarkably well. The dam just alluded to, at the Fairmount waterworks, withstood a great flood which occurred at the breaking up of the ice, on the 21st February 1832, without sustaining the smallest injury. On that occasion the water of the Schuylkill flowed over the top of the dam in a solid body no less than eight feet eleven inches in depth. As the erection of the dam impeded the navigation of the river, the Water Company had to compensate the Schuylkill Navigation Company by forming a canal, marked *h h* in Plate XIII., for the passage of their coal barges. This canal is about 900 feet in length. It has two locks of six feet lift each, and one guard-lock at the upper extremity.

The water is admitted into the mill-race *d*, by three archways at *c*, which have a water-way sixty-eight feet in breadth, and, when the river is in its ordinary state, admit a body of water six feet in depth. These archways can be shut by means of gates, and the whole of the water can be drawn off from the mill-race, if required, by opening a sluice communicating with the part of the river below the dam. The mill-race, which is excavated in solid rock, was a most laborious and expensive work. It is 419 feet in length, and 140 feet in breadth ; its depth varies from sixteen to sixty feet.

From *d*, the water flowing through the wheel-houses *e e*, drives the water-wheels, and afterwards makes its escape into the Schuylkill. The wheel-houses have been built of a sufficient size to admit of eight wheels and eight force-pumps being employed to raise the water In 1837 only six of the wheels and six force-pumps had been put up. The average daily quantity of water raised by each pump during the last year, was 530,000 gallons, and the whole quantity of water distributed from the reservoirs per day, to 19,678 householders, was 3,122,664 gallons. It has been calculated that thirty gallons of water, acting on the wheel, raised one gallon into the reservoir.

The water-wheels vary from fifteen to sixteen feet in diameter. They are fifteen feet in breadth, and make thirteen revolutions per minute. The spokes, rims, and buckets are formed of wood, but they revolve on cast-iron axles, weighing five tons each. The working of the wheels is impeded during spring tides, by the water rising upon them ; but it has been found that their motion is not affected until the back-water rises about sixteen inches on the wheel. They are stopped, however, on an average, about sixty-four hours every month from this cause.

The pumps are common double-acting force-pumps, having a stroke of six feet, worked by cranks attached to the axles of the paddle-wheels. The height to which the water is forced, is no less than ninety-two feet, and the most substantial work is necessary to insure the stability of the pumping apparatus, under the pressure of a column of water of so great a height.

A cast-iron main, sixteen inches in diameter, leads from each of the force-pumps to the reservoirs. The communication between the force-pumps and the reservoirs, can be cut off by a stop-cock, placed on the main, so that, when the pumps are not in motion, they can be relieved from the pressure of the column of water. The shortest main is 284 feet in length.

The reservoirs for containing the water are placed at an elevation of 102 feet above the level of low water, and fifty-six feet above the highest part of the streets of Philadelphia. There are four reservoirs, the aggregate area of which is about six acres. The reservoirs are founded on an elevated rock, but the water is retained by means of artificial walls and embankments. The side walls of the reservoirs are built with stone, behind which there is a backing of clay puddle, two feet in thickness, and the whole is surrounded towards the outside, by an embankment of earth, sloping at the rate of one perpendicular to one horizontal, and covered with grass sods. The reservoirs are paved with bricks, laid with lime-mortar, on a layer of clay-puddle, and well grouted, to prevent leakage. The depth of water in the reservoirs is twelve feet three inches, and when filled, they contain upwards of twenty-two millions of gallons of water. There is a considerable advantage in having four reservoirs. The water, after being discharged from the force-pumps into one of them, passes through all the other reservoirs, between each of which there is a filter, so that any impurities in the water are extracted during its passage from one cistern to another, and prevented from entering the pipes, which distribute it to the town.

The water is conveyed from the reservoirs, and distributed through the town, in 98¾ miles of cast-iron pipes. About one-half of these pipes was cast in America, and the remainder were imported from this country. The two mains leading from the reservoirs to the town, measure twenty-two inches in diameter. The small mains and pipes which have been laid in the streets, measure from three to twelve inches in diameter. The pipes are formed in the usual manner, and the different lengths are connected by spigot and faucet joints. The average cost of the whole of the pipes and mains laid down, was 7s. 1½d. per lineal foot.

The very small cost at which the town is now supplied, is an ample ground for having substituted, even at considerable outlay in the first instance, the system of raising by means of water, instead of steam power; steam having been used at the Fairmount works previous to the year 1822. The expenditure, including repairs and salaries connected with the works, for distributing a daily supply of 3,122,664 gallons of water, was, in 1836, L.2800. The following information regarding the details of this most interesting and efficient work, are drawn up by Mr Graffe, the superintendent, and printed in the Water Company's annual report for the year 1836:—

	Gallons.	Gallons.
The Reservoir No. 1. was finished in 1815, and contains, ...	3,917,659	
The Reservoir No. 2. was finished in 1821, and contains, ...	3,296,434	
The Reservoir No. 3. was finished in 1827, and contains, ...	2,707,295	
Containing, . .		9,921,388
The first section of Reservoir No. 4. was finished in 1835, and contains, ...	3,658,016	
The second section of Reservoir No. 4. was finished in 1836, and contains, . . .	4,381,322	
The third section of Reservoir No. 4. was finished in 1836, and contains, . . .	4,071,250	
		12,110,588
The Reservoirs contain together, .		22,031,976

Reservoir No. 1. cost, ...	D.32,508.52
Reservoir No. 2. cost, ...	9,579.47
Reservoir No. 3. cost, ...	24,521.75
First, second, and third sections of Reservoir No. 4. cost,	67,214.68
Total, .	D.133,824.42

The whole expense of the reservoirs amounted to 133,824 dollars, which is equal to about L.26,765.

"The water of the reservoirs covers a surface exceeding six acres. The reservoirs are each 12 feet 3 inches deep, and are elevated above the water in the dam 96 feet perpendicular.

"The water flowing from the reservoirs for the supply of the city and districts, per day, at different periods of the year 1836, was as follows :—

	Gallons.
From February 1st to the 21st, in very cold weather, .	1,769,800
... February 21st to March 20th, ...	2,113,257
... March 20th to June 3d, ...	3,046,120
... June 3d to July 22d, ...	3,942,643
... July 22d to September 9th, ...	4,152,917
... September 9th to October 28th, . . .	3,679,800
... October 28th to December 31st, . . .	3,154,114

"The average daily supply, in 1836, was 3,122,664 gallons. The above supply of water is distributed to 16,678 tenants by private pipes, and to 3000 families by public pumps, making the total number of families supplied 19,678.

"The quantity of iron pipes laid for the distribution of the water is as follows :—

	Miles.
In the city, ...	58
In the district of Spring Gardens, ...	$11\frac{5}{8}$
In Southwark, ...	$10\frac{1}{4}$
In the Northern Liberties, ...	$12\frac{7}{8}$
In Moyamensing, ...	$2\frac{5}{8}$
In Kensington, ...	3
Together,	$98\frac{3}{4}$

"The water rents collected for the year 1837 are as follows:—

In the city, ...	D.57,080.50
Including rents on the Girard estate, and rents due by H. J. Williams and others at Fairmount, .	1,048.50
In Spring Gardens, ...	13,674.25
In Southwark, ...	10,517.50
In the Northern Liberties, ...	20,009.37
In Moyamensing, ...	1,956.00
In Kensington, ...	2,146.25
Total,	D.106,432.37

Amounting, in all, to about 106,432 dollars, which is equal to about L.21,286.

"The expenses for the water-power works connected with the applicable parts of the former steam-works, were, December 31. 1831, ...	D.1,138,323.54
Add the expenses for reservoirs, iron pipes, &c. in 1832, ...	65,195.58
Do. Do. in 1833,	37,354.06
Do. Do. in 1834,	65,163.36
Do. Do. in 1835,	73,288.38
Do. Do. in 1836,	71,706.51
	D.1,451,031.43
From which deduct for the support of working machinery, materials, salaries, &c. 14,000.00 dollars per annum for the last five years, . .	70,000.00
Leaves the expenditure for the permanent works, up to 31st December 1836, ...	D.1,381,031.43

The expenditure for permanent works, therefore, amounts to 1,381,031 dollars, which is equal to about L.276,206.

The supply of water for the town of Richmond in Virginia, is procured from the James River, in the same manner as at Philadelphia; but the works are on a much smaller scale. The water is raised 160 feet by two water-wheels into two reservoirs, measuring 194 feet in length, 104 feet in breadth, and ten feet eight inches in depth, which are capable of containing upwards of two millions of gallons of water. Before leaving the reservoirs, the water is purified by passing through two gravel filters. The water-wheels are eighteen feet in diameter, and ten feet in breadth, and the fall is ten feet. The barrels of the two force-pumps are nine inches in diameter, and six feet in length of stroke, and, in the ordinary state of working, when only one wheel is in operation, raise about 400,000 gallons of water in twenty-four hours.

The cast-iron main which leads from the pumps to the reservoir is eight inches in diameter and about 2400 feet in length. Mr Stein was engineer for the work, which is said to have cost about L.20,000.

Pittsburg, on the Ohio in the State of Pennsylvania, is supplied with water from the river Alleghany. It is raised by a steam-engine of 84 horses power into a reservoir capable of containing 1,000,000 gallons of water, and elevated 116 feet above the level of the river. The main leading from the pumps to the reservoir is fifteen inches in diameter, and the pump raises 1,500,000 gallons in twenty-four hours.

Montreal also is supplied in the same manner from the water of the St Lawrence, which is raised by steam power to an elevated reservoir, and then distributed through the town.

The following account of the water-works which have lately been established at Cincinnati, on the Ohio in the State of Ohio, is given by Mr Davies the Superintendent.

"The Cincinnati water-works were constructed in 1820. The water was taken from the Ohio river, by a common force-pump, worked by horse-power, placed upon the bank of the river, sufficiently near low water-mark to be within the usual atmospheric pressure, and thrown from that point to the reservoir, 160 feet above low water-mark, from which it was conveyed to the town in wooden pipes. The town at that time afforded no inducement for a larger supply of water than could be brought through wooden pipes of three inches and a half in diameter, consequently the works at the river were only calculated to supply a pipe of that size. Only a short time, however, was necessary, to prove the necessity of an increase, and a change from horse power to steam.

"The works now consist of two engines, one propelling a double force-pump of ten inches in diameter, and four feet stroke, throwing into the reservoir about 1000 gallons a minute; the other propelling a pump of twenty inches in diameter, eight feet stroke, and discharging about 1200 gallons per minute. The reservoirs are built of common limestone; the walls are from three to six feet thick, and grouted. The water is conveyed immediately to the town, without being permitted to stand or filter. Iron pipes of eight inches in diameter convey it through the heart of the town, from which it branches in wooden pipes of from one and a half to three and a half inches in diameter. From these it is conveyed into private dwellings in leaden pipes at the expense of the inhabitants, who pay from eight to twelve dollars* per annum, according to the purposes for which it is used. Each family, of course, use any quantity they choose, their hydrants communicating freely with the main-pipes. The iron-pipes are made in lengths of nine feet each, and connected together by the spigot and faucet joint run with lead, which occupies a space round the pipe of three-eighths or half an inch in thickness."

Albany on the Hudson is principally supplied with water procured in the high ground in the neighbourhood, and conveyed in a six-inch pipe for a distance of about three miles to a reservoir near the town.

Troy, on the eastern or left bank of the Hudson, about fourteen miles above Albany, is also abundantly supplied with good water collected in the high ground in the neighbourhood. The reservoir stands about one-third of a mile from the town, and is seventy feet above the level of the streets. It is capable of containing 1,900,000 gallons, and the water is conveyed from it to the town in a main twelve inches in diame-

* From about L.1, 12s. to L.2, 8s.

ter. The works are said to have cost L.23,000. The annual expense of conducting them is L.160.

The only supply of water which the inhabitants of New York at present enjoy is obtained from wells sunk in different parts of the town. The water is raised from these wells by steam-power to elevated reservoirs, and thence distributed in pipes to different parts of the town. Some of the wells in New York belong to the Manhatten Water Company, and some to the corporation. One well, belonging to the corporation, is 113 feet in depth. For the purpose of collecting water, there are three horizontal passages leading from the bottom of the well, which measure four feet in width, and six feet in height; two of them are seventy-five, and the third is one hundred feet in length. This well cost about L.11,500, and yields 21,000 gallons in twenty-four hours. There are many other wells in the town, some of which are said to produce 120,000 gallons in twenty-four hours. This mode of collecting water in subterraneous galleries has been successfully practised in this country, on a great scale, at the water-works of Liverpool, by Mr Grahame, the engineer to the Harrington Water Company. The supply at New York is far from being adequate to the wants of the inhabitants; and the water in most of the wells being hard and brackish, is not suitable for domestic purposes.

New York is built on a flat island, which is nearly surrounded by salt water, so that the method that has been resorted to for the supply of Philadelphia and most other towns in the United States is altogether impracticable in that situation. Many plans have been proposed, and, among others, that of throwing a dam across the Hudson, so as to exclude the salt water; but as a free passage, by means of locks, must be preserved for the numerous vessels which navigate the river, the success of such a plan seems very doubtful.

Many engineers in the United States, of great reputation, have made surveys of the country in the neighbourhood of New York, in order to devise a plan for the supply of the city with water, and they have proposed to effect this object, so important to its inhabitants, by conveying the water of the river Croton in a tunnel to New York. The point from which the water is intended to be withdrawn, is about thirty miles distant from the city. The estimate for the entire execution of the work, is upwards of one million Sterling.

The situation of Boston is somewhat like that of New York. It is surrounded by the sea, and the supply of good water is far from being sufficient for the inhabitants. Mr Baldwin, civil engineer, has made a survey and plan for the supply of the town, in which he contemplates bringing water from some springs in the neighbourhood.*

* Report on introducing pure water into the City of Boston. By Loammi Baldwin, C. E. Boston, 1835.

At present the town is supplied chiefly from wells. According to Mr Baldwin's report, there are no less than 2767 wells in Boston, thirty-three of which are Artesian. Only seven, however, out of the whole number, produce soft water; and of these, two are Artesian.

Great difficulty has been experienced in forming many of the wells on the peninsula of Boston, in some of which, on tapping the lower strata, the water is said to have risen to seventy-five, or eighty feet above the level of the sea.*

The following very interesting remarks regarding two of these wells, are quoted by Mr Storrow in his Treatise on Water-works.

"Dr Lathrop gives the following history of a well dug near Boston Neck.† 'Where the ground was opened, the elevation is not more than one foot, or one foot and a half above the sea at high water. The well was made very large. After digging about 22 feet in a body of clay, the workmen prepared for boring. At the depth of 108 or 110 feet the augur was impeded by a hard substance; this was no sooner broken through and the augur taken out, than the water was forced up with a loud noise, and rose to the top of the well. After the first effort of the long confined elastic air was expended, the water subsided about six feet

* A Treatise on Water-works. By Charles S. Storrow. Boston, 1835.

† Memoirs of the American Academy of Arts and Sciences, vol. 3.

from the surface, and there, remains at all seasons ebbing and flowing a little with the tides.'

"Dr Lathrop observes, that the proprietors of this well were led to exercise great caution in carrying on the work, by an accident which had happened in their immediate neighbourhood. 'A few years before, an attempt was made to dig a well a few rods (16½ feet) to the east near the sea. Having dug about 60 feet in a body of clay without finding water, preparation was made in the usual way for boring; and, after passing about 40 feet in the same body of clay, the augur was impeded by stone. A few strokes with a drill broke through the slate covering, and the water gushed out with such rapidity and force, that the workmen with difficulty were saved from death. The water rose to the top of the well and ran over for some time. The force was such as to bring up a large quantity of fine sand, by which the well was filled up many feet. The workmen left behind all their tools, which were buried in the sand, and all their labour was lost. The body of water which is constantly passing under the immense body of clay, which is found in all the low parts of the peninsula, and which forms the basin of the harbour, must have its source in the interior, and is pushed on with great force from ponds and lakes in the elevated parts of the country. Whenever vent is given to any of those subterranean currents, the water will rise, if it have opportunity, to the level of its source.'"

CHAPTER XI.

LIGHTHOUSES.

Parts of the United States in which Lighthouses have been erected—Great extent of coast under the superintendence of the Lighthouse Establishment—The uncultivated state of a great part of the country, and the attacks of Indians a bar to the establishment of Lights on the coast—Introduction of Sea Lights in America—Description of the present establishment—Number of Lighthouses, Floating Lights, and Buoys—Annual Expenditure—Management — Superintendents — Light-Keepers — Supplies of Stores, &c.—Lighting Apparatus—Distinctions of Lights—Communication on the subject from Stephen Pleasonton, Esq., Fifth Auditor of the Treasury.

THE parts of the territory of the United States on which lights have been erected under the management of the General Lighthouse Establishment, are, *First*, The eastern coast of the country from Passamaquoddy Bay, the boundary between the American and British dominions, to the State of Texas, in the Gulf of Mexico, a stretch of coast extending to upwards of 3000 miles, exposed to the Atlantic Ocean. *Second*, The courses of the rivers Mississippi and Ohio, extending to about 1250 miles. *Third*, The southern shores of Lakes Ontario, Erie, Huron, and Michigan, including a line of coast of not less than 1200 miles in extent. In addition to these great outlines, lights have also been placed on some of the smaller rivers and lakes, for the purpose of facilitating their navigation.

The western coast of the country, which is washed by the Pacific Ocean, is entirely cut off from any communication with the inhabitants of the United States by a great tract of uncultivated and unexplored land, stretching from the northern to the southern extremity, and flanked by the rugged ridges of the Rocky Mountains. The United States of America, therefore, are quite unapproachable from the Pacific. The western coast of the country (a great part of which has never been explored), is still far removed from the limits of civilization, and is inhabited only by tribes of wandering Indians.

The whole extent of coast under the jurisdiction of the American Lighthouse Establishment embraces the three compartments which have been enumerated, and is not less than 5450 miles, while the coast of Great Britain and Ireland may be stated at 2800 miles, and that of France at 1100 miles. The uninhabited and desolate condition of a large part of the coast proves a great bar to the regular and efficient establishment of lighthouses. This fact has been strikingly exemplified, and its consequences severely felt, in the State of Florida, which is said to be the most dangerous coast in the United States of North America. The country in this State is almost wholly uncultivated. It is still in many places peopled only by remnants of Indian tribes, who have shewn their hostility to the introduction of any thing like civilization, by opposing the erection of lighthouses on the coast, and in some places, by burning the lighthouse towers, and even murdering the keepers. In one instance, a light-keeper on the coast of Florida, after defending himself for a considerable time against an attack made by a body of Indians, was at last forced to take refuge in the balcony of the lighthouse, where he was shot by the arrows of the assailants. The following extract, taken from a letter addressed by the Fifth Auditor, to the Secretary of the Treasury of the United States, shews the difficulty that is often encountered in transacting the business of the lighthouse establishment :—" A contract was made in the month of July last, for rebuilding the lighthouse at Cape Florida, and the contractor proceeded to that place with materials and men to execute the work; but finding that hostile Indians were in the neighbourhood, he returned to Boston (a distance of about 1300 miles) without effecting his object. When the contract was made, there was just reason to believe that the Indian war was at an end, and that the work could be done with safety."

The fact of a lighthouse system having been extended to the remotest corners of so extensive a coast, under circumstances so inauspicious and unfavourable, is what could hardly have been looked for, and is certainly highly creditable to the government of the United States and to the officers of the Lighthouse Establishment. Even the most superficial observer cannot fail to discover that there is a striking contrast between the regulation of that establishment and the efficient and admirable systems pursued by the Lighthouse Boards of Great Britain and France; but a candid enquirer will rather be disposed to admire the activity and zeal which have extended the benefit of lighthouses to remote and unhospitable regions, of difficult access, than to wonder at the defects of the system which has been established for the purpose of carrying that important object into effect.

The period at which lighthouses were first used for facilitating navigation is not correctly known. The Pharos of Alexandria seems to have existed as early as 300 B.C. In England they were in use in the reign of Henry VIII.; in Scotland in the reign of James VI.; and in Ireland in the reign of George II. We are perhaps indebted to France for the introduction of a more perfect system of management, the Government of that country having first placed the management of the lighthouses under the charge of Engineers.

The date at which the first sea light was exhibited on the coast of America is not exactly known; but the management of the lighthouses appears to have been undertaken by the Government of the United States, and a system for conducting them regularly organized in the year 1791, at which period they were only ten in number. These appear to have been erected in the States of Massachusetts, New York, and Virginia, which were the earliest settlements in the country. The whole number of lighthouses, including harbour lights (which are also under the control of the General Lighthouse Board), in 1837, was 202. Of these about 172 are situated on the sea coast, and the remaining 30 are on the great lakes and rivers. There are also 26 floating light ships, which are moored in the vicinity of particular dangers on the coast, and vary in size from 50 to 225 tons register, according to their position and importance. Their lights are exhibited in the usual manner from lanterns suspended at the mast-heads of the vessels. In addition to the duties connected with the management of the lights, the Board has also the charge of upwards of 600 buoys and beacons placed on different parts of the coast.

The total expenditure connected with the lighthouse establishment of America for the year 1837 was 356,863 dollars, which is equal to about L.71,352. Of this outlay the sum of L.19,652 was expended in paying the " salaries of principal officers, superintendents, and light-keepers ; L.17,720 in the purchase of oil and other stores for the lights, and in repairing lamps ; L.7000 in supporting the buoys ; L.13,000 in keeping the light ships in repair, and L.13,980 in repairing lighthouse towers and executing new works.

The business of the Lighthouse Establishment, as has already been noticed, is under the immediate control and management of the Government. The official person to whom the duties of this department have been specially assigned, is the Fifth Auditor of the Treasury, who superintends the building and maintenance of the various lighthouses, floating-light ships and buoys on the coast, and the general expenditure connected with the establishment. He resides at Washington, the seat of Government of the United States, and does not himself visit the lighthouse stations, but conducts the business with the assistance of superintendents. This vast stretch of shore is divided into forty-one districts, over each of which a superintendent is placed, for the discharge of the coast duty. The person chosen to fill this office is generally resident in the part of the country where his duty lies. Some of these superintendents have as many as twenty-four lighthouses, while others, in parts of the country where the lights are few in number and widely separated, have proportionally fewer under their charge. The duty of the superintendent consists in visiting and inspecting the lighthouses of his district, reporting the repairs required on them, and seeing the same executed, and in receiving from the keepers of the lighthouses quarterly returns of the quantity of the stores expended. These he transmits to the Fifth Auditor of the Treasury. The superintendent also makes an annual report on the general state of the lighthouses, and the conduct of the light-keepers under his charge. This officer is paid for his services at the rate of two and a half per cent. on the amount of his annual disbursements, a mode of remuneration which appears to be of very questionable propriety.

One keeper only is appointed to the charge of each lighthouse, who, as before noticed, makes a quarterly return to the superintendent of the qu at ity of stores expended, but keeps no journal of the times at which the lamps are lighted and extinguished, and no register of the weather. The keepers' salaries range from L.50 to L.120 per annum, according to the favourable or unfavourable nature of the situation at which they are placed, and keepers of the floating-light ships are paid at nearly the same rate. The desolate and uninhabited state of many of the situations in which the lights are placed, as well as the fact of there being only one responsible person at each station, render it difficult to conceive how the duties of the light-keepers can be efficiently performed; while the imperfect nature of their periodical reports, and the remote intervals at which they are made, afford very little security for, or at all events satisfactory evidence of, the fulfilment of the important duties committed to them, on the faithful discharge of which, the lives and fortunes of many individuals must constantly depend.

The furnishing of oil and other stores, and the repairs necessary for keeping the lamps in a proper working state, as well as the delivery of the supplies at the different stations, are let at a gross sum by contract for five years at a time. In 1837, this contract was executed at 35 dollars 87 cents, or L.7 : 9 : 5 per lamp per annum, a sum which, taking into consideration the actual value of the oil and supplies consumed, and the difficulty and expense of delivering them, seems quite inadequate. The contractor is also expressly bound, on landing the supplies, to examine the state of the several lighthouses, and send an annual report to the Fifth Auditor of the Treasury, specifying the repairs on the light-towers or dwelling-houses, which he considers necessary for maintaining the efficiency of the lights. This arrangement is understood to have been adopted as a check on the conduct of the superintendents.

The apparatus used in illuminating the American lighthouses is in general constructed on the catoptric principle. The reflectors in use are made of polished tin-plate, and measure from nine to eighteen inches in diameter. They are inferior to those employed on the coasts of Great Britain and France, which are of much larger dimensions, and made of copper plated with silver, and highly polished. The common argand lamp, similar to that in use in British lighthouses, but of a smaller size, is employed in illuminating the reflectors. Spermaceti oil, the produce of their South Sea fishery, is burned in all the lighthouses. Some experiments have lately been made with oil produced from cotton-seed, which have been considered satisfactory, and it is expected that ere long this description of oil will be generally adopted for lighting the American coast. Common crown-glass is used for the windows of the lighthouses, while in this country polished plate-glass, which, from its greater strength and purity, is much better suited for the purpose, is universally employed. The characteristics used in the American lighthouses, for the distinction of one light from another, are the stationary, revolving, red, and double lights. On the British coasts, seven different distinctive lights have been introduced, with much success, in those lighthouses which are illuminated on the catoptric principle.

These seven distinctions are called ***stationary, revolving white, revolving red and white, flashing, intermittent, double, and leading lights.*** The first exhibits a steady and uniform appearance, and the reflectors used for it are of smaller dimensions than those employed in lights which revolve. This is necessary in order to permit them to be ranged round the circular frame, with their axes inclined at such an angle as shall enable them to illuminate every point of the horizon. The revolving light is produced by the motion of a frame with three or four faces, having reflectors placed on each of its sides ; and as the revolution exhibits a light gradually increasing to full strength, and in the same gradual manner diminishing to total darkness, its appearance is extremely marked and obvious to the mariner. The alternation of red and white lights, is produced by the revolution of a frame, whose alternate faces present red and white lights. The flashing light is produced in the same manner as the revolving light, but owing to a different construction of the frame, and the greater quickness of the revolution, a totally different and very splendid distinction is obtained. The lightest and darkest periods being but momentary, this light is characterized by a rapid succession of bright flashes, from which it gets its name. The intermittent light is distinguished by its bursting suddenly on the view, and continuing steady for a short time, after which it is suddenly eclipsed for half a minute. This striking appearance is produced by the perpendicular motion of circular shades in front of the reflectors, by which the light is alternately hid and displayed. The double light consists of two lights exhibited from the same tower, the one raised above the other. The leading lights are exhibited from two towers, one higher than the other, and when seen in one line, they form a direction for the course of shipping.

To those acquainted with the British lighthouse system, the remarks that have been made regarding the general management and the details of the American lighthouses, will shew that much may still be done in improving this important class of public works in that country; and it is to be hoped that when the hour of improvement arrives, a rapid stride will be made, so as at once to bring into force all the best attainments which have been effected in Europe. The dioptric instruments of Fresnel are now generally acknowledged, under certain circumstances, to increase the power, while they lessen the expense of illuminating lighthouses, and have lately been introduced into this country, under the directions of the Commissioners of the Northern Lights, by Mr Alan Stevenson. The last work of this kind is the employment, at the fixed light of the Isle of May, of refractors formed into a cylindric zone or belt, instead of the polygon used in lights of the first order in France. This is, in fact, merely the extension of the dioptric zones of Fresnel's harbour-light apparatus, to the scale of a great cylinder six feet in diameter. It is to be hoped the Americans will at once adopt the dioptric system, at least in all lighthouses whose situation is such as to insure a constant and efficient superintendence of the duty and conduct of the light-keepers. This limitation of this admirable system seems necessary, from the greater care required in watching a dioptric light, which is illuminated by means of a mechanical lamp, somewhat delicate in its movements, and easily deranged; so that wherever the light-keepers are left for a long time without an inquiry into the manner in which they discharge their duty, perhaps the catoptric system in its most improved state is best calculated to insure, if not entire efficiency, at least greater confidence in the light.

Since writing the foregoing pages, I had the honour

to receive the following communication from Stephen Pleasonton, Esq., in answer to some inquiries made by me relative to the revenue by which the American lights are supported.

" *Treasury Department, Fifth Auditor's Office,*
May 1. 1838.

" DEAR SIR,—I had the honour to receive your letter of the 22d March, yesterday; and it is with great pleasure that I now furnish the information desired in relation to our Light House Establishment.

" The lighthouses of the United States are built, fitted up, and kept in operation, by appropriations made by Congress from the general funds of the government. There is no tax imposed upon commerce, upon the States, or upon individuals, for this purpose. The whole expense is paid from the revenue provided for the support of the Government generally.

" After lighthouses are built, for which special appropriations are made, appropriations are annually made for supporting the establishment.

" That for the year 1838 is as follows :—

	Dollars.
" For the support of lighthouses, floating lights, beacons, and buoys, supplying lighthouses with oil, tube-glasses, buff-skins and whiting, and keeping the apparatus in repair, viz.: 2215 lamps ($35\frac{87}{100}$ dollars per lamp),	88,600
Salaries of 202 keepers of lighthouses, . . .	80,113
Salaries of 27 keepers of floating lights, . . .	14,150
Weighing, mooring, cleansing, repairing, and supplying the loss of beacons, buoys, chains, and sinkers, .	35,000
Incidental expenses, repairs and improvements to lighthouses and the buildings connected therewith,	70,000
Carry forward,	287,863

	Dollars.
Brought forward,	287,863
Incidental expenses, seamen's wages, repairs and supplies to floating lights, ...	65,000
Expenses of a Board of officers in examining and reporting the conditions of all the lighthouses annually,	4,000
	356,863

(356,863 dollars, or about L.71,352.)

"The lighthouses of this country are supplied by contract, to continue in force for five years at a time, with oil, tube-glasses, buff-skins, whiting, diamonds for cutting glass, and every other thing necessary to keep them lit, the contractor also keeping all the apparatus in complete repair, for the sum of $35\frac{87}{100}$ dollars (L.7 : 9 : 5) per lamp, annually. If the oil is found not to be good on trial (for we have found no other way of testing it), he takes that away, and supplies that which is good.

"The repairs of the buildings are a separate charge, and are made by direction of this office.

"Besides the distinctions of fixed and revolving lights, we have two other modes of distinguishing our lighthouses from each other. The one is by producing a deep red light, which is done by employing red tube-glasses; and the other is by placing one light above another in the same tower, leaving a space of several feet between them.

"I have the honour to be, very respectfully, Sir, your obedient Servant,

S. PLEASONTON,
Fifth Auditor of the Treasury,
and Acting Commissioner of the Revenue.

"David Stevenson, Esq. Civil Engineer, Edinburgh."

CHAPTER XII.

HOUSE-MOVING.

THE lowest wages in the United States for labourers employed at railways or canals, in 1837, were one dollar or about 4s. 2d. a-day, while the same class of workmen in this country receive 2s. per day. In consequence of the great value of labour, the Americans adopt, with a view to economy, many mechanical expedients, which, in the eyes of British engineers, seem very extraordinary.

Perhaps the most curious of these, is the operation of moving houses, which is often practised in New York. Most of the old streets in that town are very narrow and tortuous, and in the course of improving them, many of the old houses were found to interfere with the new lines of street, but instead of taking down and rebuilding those tenements, the ingenious inhabitants have recourse to the more simple method of moving the whole *en masse*, to a new site. This was, at first, only attempted with houses formed of wooden framework, but now the same liberty is taken with those built of brick. I saw the operation put in practice on a brick house, at No. 130 Chatham Street, New York, and was so much interested in the success of this hazardous process, that I delayed my departure from New York for three days, in order to see it completed. The house measured fifty feet in depth by twenty-five feet in breadth of front, and consisted of four storeys, two above the ground-floor, and a garret-storey at the top, the whole being surmounted by large chimney stacks. This house, in order to make room for a new line of street, was moved back fourteen feet six inches from the line which the front wall of the house originally occupied, and as the operation was curious, and exceedingly interesting in an engineering point of view, I shall endeavour, by referring to the accompanying diagrams, to describe the manner in which it was accomplished. Fig. 1 is an elevation of the gable, and Fig. 2 an elevation of the front of the house.

Fig. 1.

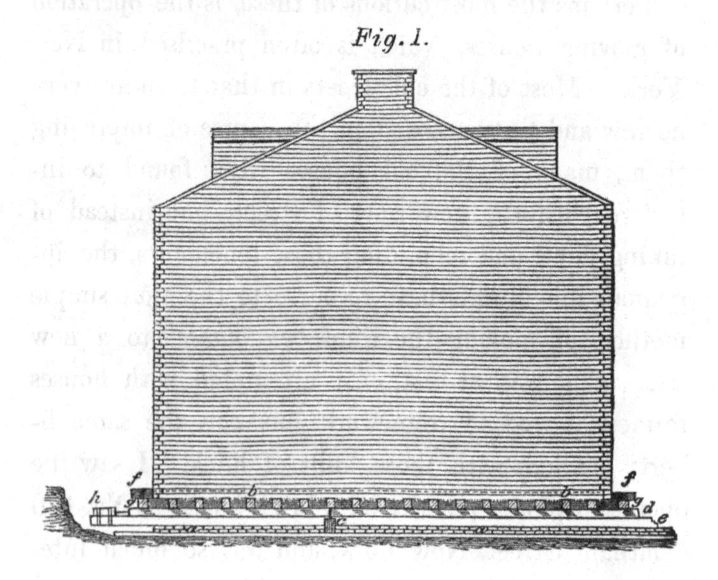

Fig. 2.

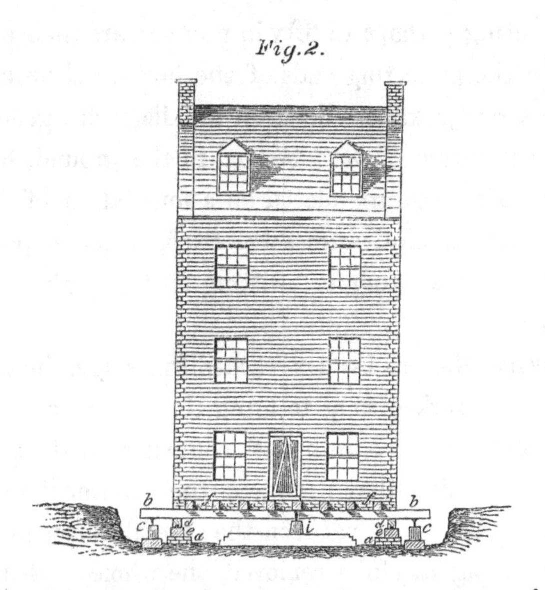

The first step in the process is to prepare a foundation for the walls on the new site which the house is intended to occupy. A trench is next cut round the outside of the house, and the lower floor being removed, the earth is excavated from the interior, so as to expose or lay bare the foundations of the side walls and gables, which are represented in the cuts by *a*. Horizontal beams of wood, marked *b*, measuring about twelve inches square, are then arranged at distances of three feet apart from centre to centre, at right angles to the direction in which the house is to be moved, their ends being allowed to project about three feet each beyond the building, through holes drifted in the gables for their reception, as shewn at *b b* in Fig. 2. A series of powerful screw-jacks, marked *c*, amounting perhaps to fifty in number, are then placed under the projecting ends of the horizontal beams, *b*. The screw-jacks, as shewn in the diagrams, generally rest on a beam of wood bedded in the ground, but in some cases they are placed on a foundation of stone. They are carefully *ranced* or fixed, so as to prevent them from kanting or twisting on the application of pressure.

When the process has reached this stage, the screw-jacks are worked so as to bring the upper sides of the horizontal beams *b*, into close contact with the gables, through which they pass, and the intermediate portions of the walls, between the several points of support, being carefully removed, the whole pressure of the gables is brought to bear on the horizontal beams *b*, which rest on the screw-jacks *c*. Two strong beams, which are represented by letters *d e* in the diagrams, are placed, one resting above the other, under each gable, (a part of which is removed for their reception) at right angles to the horizontal beams *b* ; the lower beam *e*, rests on the old foundation of the house, which is levelled for its reception, and the upper beam *d*, is firmly fixed, by means of cleats of wood and spikes, to the horizontal beams *b*, passing through the house. The lower beams form the road, as it were, on which the upper ones, supporting the house, slide. The lower beams are accordingly extended, as shewn at *e*, Fig. 1, by means of similar beams, resting on a firm foundation, to the new site of the house. After the beams, *d e*, have been securely placed close under the horizontal beams, *b*, the screw-jacks are unscrewed, and the whole weight of the gables is again made to bear on the foundations. Holes, at distances of about three feet apart from centre to centre, are next drifted in the front and back walls of the house, through which logs, marked *f*, are inserted, in the same way as formerly described in the gables. The ends of these logs project about three feet beyond the faces of the walls, and are supported by cross beams, shewn at *g g*, Fig. 1, the ends of which, rest upon the beams, *d*, under the gables. The intermediate portions of the front and back walls, between the supporting beams, being removed in the same manner as the gables, the whole weight of the building rests on the lower beams, *d* and *e*, on which the motion is to take place. A very powerful screw-jack, shewn at *h*, Fig. 1, is fixed, in a horizontal position, to each of the beams, *e*, on which the house is to move. The ends of the screw-jacks butt against the upper beams, *d*; and when they are worked, the upper beams, bearing the whole weight of the house, slide smoothly along on the lower beams, *e*. The two beams are well greased; and a groove in the upper, and a corresponding feather on the surface of the lower one, insure a motion in the direction of their length. The length of the screws in the screw-jacks, *h*, is about two feet; so that if the house has to be removed to a greater distance than that included in their range, they are unfastened, and again fixed to the beam, *e*, when the house is then propelled other two feet. In this way, by prolonging the beams *e*, and removing the screw-jacks, the house may be moved to an indefinite distance.

When the house has been brought directly over the foundation which was prepared for it, and which we shall now suppose to be represented by *a* in the cuts, the spaces between the beams *f* and the foundation *a*, in the front and back walls of the house, are built up, and also the intermediate spaces between the several beams *f*. Screw-jacks, as shewn at *c* and *i*, are then ranged all round the house under the ends of the projecting beams; they are now, as formerly, placed on firm foundations, and properly braced, to prevent them from twisting or kanting. These screw-jacks are then all worked, and the weight of the house is transferred to them from the beams *d*, *e*, *g*, which are carefully removed. The space between *a a*, Fig. 1, and the horizontal beams *b*, which was occupied by the beams *d*, *e*, is now built up, and also the intermediate spaces between the beams *b*. The screw-jacks *c* are then slackened one after another, and the beams *b* withdrawn, the space which each occupied being carefully built up before another screw-jack is removed. The same process is performed with the beams *f*, and the house then rests on its new foundation *a*, which, in the case I saw in New York, was fourteen feet six inches from the spot on which the house was built.

The operation I have attempted to describe is attended with very great risk, and much caution is necessary to prevent accidents. Its success depends chiefly upon getting a solid and unyielding base for supporting the screw-jacks *c i*, and for the prolongation of the beam *e* to the new site which the house is to occupy. It is further of the utmost importance that in working the screws their motion should be simultaneous, which in a range of 40 or 50 screw-jacks is not very easily attained. The operation of drifting the holes through the walls also requires caution, as well as that of removing the intermediate pieces between the beams *b* and *f*, which pass through both walls. The space between the beams is only two feet, and the place of the materials removed, is, if necessary, supplied while the house is in the act of moving, by a block of wood which rests on the beams *d*. The screw-jacks *h*, by which the motion is produced, require also to be worked with the greatest caution, as the cracking of the walls would be the inevitable consequence of their advancing unequally.

Notwithstanding the great difficulty attending the successful performance of this operation, it is practised in New York without creating the least alarm in the inhabitants of the houses, who, in some cases, do not even remove their furniture while the process is going forward. The lower part of the house which I saw moved was occupied as a carver and gilder's shop; and on Mr Brown, under whose directions the operation was proceeding, conducting me to the upper storey, that he might convince me that there were no rents in the walls or ceilings of the rooms, I was astonished to find one of them filled with picture frames and plates of mirror glass, which had never been removed from the house. The value of the mirror glass, according to Mr Brown, was not less than 1500 dollars, which is equal to about L.300 Sterling; and so much confidence did the owner of the house place in the success and safety of the operation, that he did not take the trouble of removing his fragile property. I understood from Mr Brown that the whole operation of removing this house, from the time of its commencement till its completion, would occupy about five weeks, but the time employed in actually moving the house fourteen feet and a half was seven hours. The sum for which he had contracted to complete the operation was 1000 dollars, which is equal to about L.200 Sterling. Mr Brown mentioned that he and his father, who was the first person who attempted to perform the operation, had followed the business of "house-movers" for fourteen years, and had removed upwards of a hundred houses, without any accident, many of which, as in the case of the one I saw, were made entirely of brick. I also visited a church in "Sixth" Street, capable, I should think, of holding from 600 to 1000 persons, with galleries and a spire, which was moved 1100 feet, but this building was composed entirely of wood, which rendered the operation much less hazardous.

NOTE ON THE MANUFACTORIES AT LOWELL.

THE manufactures of the United States are becoming every day more important. The largest factories in that country have been established at Lowell, on the banks of the Merrimac, in the State of Massachusetts. The following statistical table, relative to the works at that place, may perhaps be useful to those interested in that subject. The first mill built at Lowell was opened in 1822 ; and in 1837, there were twenty-seven mills in the town, which employed no fewer than 7912 persons. The machinery in all the mills which I had an opportunity of visiting at Lowell, was excellent. In the cotton-mills, in particular, the carding-machines and spinning-frames were very highly finished ; and the dressing-machines were more simple, and apparently quite as effective as any I have ever seen in this country. With the exception of the works of Mr Smith at Deanston, I have seen no establishment in which the beneficial effects of good machinery and excellent regulation were more obvious than at the Lowell works in the United States.

Pittsburg is also entirely a manufacturing town; and, in addition to several cotton-mills which have been built in it, there are several glass-works and iron-foundries. The sandstone from which the glass is made is found on the banks of the Alleghany river, about 100 miles from Pittsburg; the ironstone is got on the banks of the Juniata and Susquehanna rivers, and brought to the town on the Pennsylvania Canal. I visited some of the glass and ironworks in Pittsburg, which are similar to those of this country; but the goods manufactured are decidedly inferior in quality.

STATISTICS OF LOWELL MANUFACTURES, January 1. 1837.

COMPILED FROM AUTHENTIC SOURCES.

Corporations.	Locks and Canals.	Merrimack.	Hamilton.	Appleton.	Lowell.	Suffolk.	Tremont.	Lawrence.	Middlesex.	Boott Cotton Mills.	Total.
Capital Stock,	600,000	1,500,000	1,000,000	500,000	500,000	450,000	500,000	1,500,000	500,000	1,000,000	8,050,000
Number of Mills,	2 shops and a smithy.	5 and Print Works.	3 and Print Works.	2	Cotton & Carpet Mill, 1 building.	2	2	5, another or bleachery preparing.	2 and Dye-House.	4, two in operation, and 2 going into operation the next season.	27, exclusive of Print-Works, &c.
Spindles,	. .	35,704	21,228	11,776	5000 cotton besides woollen.	11,264	11,520	31,000	4620	14,016	146,128
Looms,	. .	1253	620	380	144 cotton 70 carpet.	352	404	910	38 broadcloth, 92 cassimere.	404	4667
Females employed,	. .	1400	860	470	375	470	460	1250	350	450	6085
Males,	500	437	230	65	200	70	70	200	185	70	1827
Yards made per week,	. .	186,000	110,000	100,000	2500 carpet, 150 Rugs, 55,000.	90,000	125,800	200,000	6300 cassimere, 1500 broadcloth.	73,000	950,250
Bales Cotton used in do.	1250 tons wrought and cast iron yearly	120	100	95	76	86	90	180	None.	60	807
Pounds Cotton wrought in do.	. .	44,000	39,000	33,000	30,000	32,000	34,000	64,000	600,000 lb. wool per ann. and 3,000,000 teasels.	21,000	297,000
Yards dyed and printed do.	. .	165,000	70,000	None.	None.	None.	None.	None.	. .	None.	235,000
Kinds of Goods made,	Machinery, Cars and Engines for Railroads.	Prints and Sheetings, No. 22 to 40.	Prints and Drillings, No. 14 to 40.	Sheetings and Shirtings, No. 14.	Carpets, Rugs, and Negro Cloth.	Drillings, No. 14.	Sheetings and Shirtings, No. 14.	Printing Cloths, Sheetings, & Shirtings, No. 14 to 30.	Broadcloths and Cassimeres.	Drillings, No. 14; Shirtings, No. 40.	
Tons Anthracite Coal per annum,	200 chaldrons smith's coal; 200 tons hard coal.	5200	2800	300	350	330	329	650	500	300	10,759
Cords of Wood per annum,	300	1500	1250		500	70	60	60	1000	70	4510
Gallons of Oil,	2300	8700	6500	3375	Olive, 4000. Sperm. 4000.	3840	3692	8217	Olive, 11,000 Sperm. 2500	3500	59,324
Diameter of Water Wheels,	13	30	13	13	13	13	13	17	17 and 12	17	
Length of do. for each mill,	14	24	42	42	60	42	42	60	46 and 21	60	
Incorporated,	1792	1822	1825	1828	1828	1830	1830	1830	1830	1835	
Commenced operations,	1822	1823	1825	1828	1828	1832	1832	1833–4	1830	1836	
How warmed,	Hot Air.	Hot Air Furnace.	Hot Air Furnace.	Hot Air Furnace.	Hot Air Furnace.	Hot Air Furnace.	Hot Air Furnace.	Steam.	Wakefield Furnace and Steam.	Hot Air.	

REMARKS TO THE FOREGOING TABLE.

Yards of cloth made per annum, ... 49,413,000

Pounds of cotton consumed, ... 15,444,000

Assuming half to be Upland, and half New Orleans and Alabama, the consumption in bales, averaging 361 lb. each, is 41,964

A pound of cotton averaging . . . . . $32\frac{1}{10}$ yards.

100 pounds of cotton will produce 89 pounds of cloth.

As regards the health of persons employed, great numbers have been interrogated, and the result shews, that six of the females out of ten enjoy better health than before being employed in the mills,—of males, one-half derive the same advantage.

As regards their moral condition and character, they are not inferior to any portion of the community.

Average wages of females, clear of board, 2.00 dlrs. per week.

of males, clear of board, . 80 cts. per day.

Medium produce of a loom on No. 14, yarn, 38 to 49 yds. per day.

... ... No. 30, 25 to 30 ...

Average per spindle, ... $11\frac{1}{10}$ yds. per day.

Persons employed by the companies are paid at the close of each month.

The average amount of wages paid per month, 106,000 dollars.

A very considerable portion of the wages is deposited in the savings bank.

Consumption of starch per annum, . . . 510,000 lb.

Consumption of flour for starch in the mills, print-works and bleachery, per annum, . . . 3,800 bushels.

Consumption of charcoal, per annum, . . 500,000 bushels.

To the above-named principal establishments may be added the extensive Powder Mills of Oliver M. Whipple, Esq.; the Lowell Bleachery; Flannel Mills; Card and Whip Factory; Planeing Machine; Reed Machine; Grist and Saw Mills;—together employing about 300 hands, and a capital of about 300,000 dollars; and in the immediate vicinity, Glass-Works, and a Furnace supplying every description of Castings.

The Locks and Canals Machine Shop, included among the twenty-seven mills, can furnish machinery complete for a mill of 5000 spindles in four months, and lumber and materials are always at command, with which to build or rebuild a mill in that time, if required. When building mills, the Locks and Canals Company employ directly and indirectly from 1000 to 1200 hands.

FINIS.

Printed in the United States
By Bookmasters